X.400
The Messaging and Interconnection
Medium for the Future

An NCC Management Report

Paul Chilton

NCC Blackwell

MANCHESTER • OXFORD

British Library Cataloguing in Publication Data

Chilton, P. A.
    X400 : the messaging and interconnection
    medium for the future
    1. Computer systems. Data transmission.
    Data elements – Standards
    I. Title
    004.6'2
    ISBN 0-85012-685-1

Published for NCC Publications by NCC Blackwell Limited.

Editorial Office: The National Computing Centre Limited, Oxford Road, Manchester M1 7ED, England.

NCC Blackwell Limited, 108 Cowley Road, Oxford OX14 1JF, England.

Typeset in 11pt Times Roman by H & H Graphics, Blackburn; and printed by Hobbs the Printers of Southampton.

ISBN 0-85012-685-1

Dedication

To my parents for their support over the years and to
Teresa for her support and patience.

Contents

Acknowledgements

I would like to thank the following people who kindly offered to review material for this publication:

Janusz Zajaczkowski, British Telecom
Robert Willmott, Independent Consultant
Bill McKinley, British Airways
Martin Parfett, NCC
Richard Wakerley, NCC
Steve Price, NCC

In addition I wish to extend my thanks to the following individuals and companies who provided a substantial amount of information for this publication:

Andrew Knowles and Sam Quinn, British Telecom
Ian Valentine and Graham Knight, Level 7 Ltd
Dave McKnight, Telecom Canada
Mike Chesher, Geisco
John Cummins, Xionics
John Williamson, INS
Rainer Jansen and Dr Harold Nottebohm, Hoechst

British Telecom	CAP
Concurrent Computer Corp Ltd	Data General
DEC	GPT (GEC Plessey Telecom)
Hewlett Packard	Honeywell Bull
IBM UK Ltd	ICL
ITL	Logica UK
McDonnell Douglas	NCR Ltd
Norsk Data Ltd	Prime Computer (UK) Ltd
Sydney Communications Ltd	Tandem
UNISYS	Wang (UK) Ltd

Finally, I would also like to acknowledge the CCITT X.400 recommendations (both the Red and Blue books) as a major source of material for this publication.

The author also wishes to acknowledge Level 7 Ltd for their help in compiling information for Chapter 7 of this book. In particular he wishes to thank Graham Knoght for his efforts in gaining information from the service providers, and Ian Valantine whose articles formed the basis of the sections on the ceBIT and Telecom 87 interworking demonstrations.

The Centre acknowledges with thanks the support provided by the Computing, Software and Communications Committee (CSC) for the project from which this publication derives.

1 The Need for Standardised Electronic Messaging

Within the world of business and commerce the need for the rapid transfer of information has always existed. Information in its varied forms is a valuable commodity which, if acquired in advance of competitors, can provide a business with its competitive edge. Evidence of the importance of such information can be gained by the rapid growth of companies whose business is based purely on information gathering and distribution, whether via paper-based (directories and circulars) or electronic (databases) means. Examples here would be companies such as Dunn & Bradstreet, Reuters, Kemps, Finsbury Data Services, Pergamon, etc.

INFORMATION TRANSFERENCE

Information is not necessarily gathered from specific sources, such as those mentioned above; rather it may originate from many different areas and by way of numerous media. Within large companies a vast amount of information is generated, the distribution of which requires co-ordination and co-operation.

Although postal services and paper-based documentation have constituted the backbone of information transfer for some considerable time now, it is apparent that electronic means are starting to have some impact. The introduction of telex, in 1932, was the first step, and its current usage and penetration is without question impressive. More recently, however, and within the last decade, there has been a gradual introduction of office automation equipment that is largely based on computer hardware.

The range of facilities covered by office automation is very broad, but in the context of information retrieval and transfer it can be narrowed to encompass the following technologies:

— telex;

— teletex;

— public electronic mail;

— private electronic mail;

— facsimile.

In the same period as the introduction of these technologies, the computer has begun to penetrate very rapidly into the business environment. Initially the computer was used to achieve very specific objectives or tasks. This meant that a computer would be the property of a department or group and it would serve their particular needs: examples here might be payroll calculation or other general data processing applications. Information storage on computers is also a major usage and it is this data which the applications will use to produce the required results.

The development of the computer since its early days has been very fast indeed. Today it is possible to buy a personal computer with the kind of memory storage capability and processing power that just over a decade earlier might only have been in the realm of a mini or mainframe installation. This rapid development has led to the distribution of computer processing power throughout the office and even down to the level of individual's desks.

Experience with computers and office automation (OA) equipment generally has quickly yielded a user requirement for their interconnection. In this way information and services, available only in small pockets, could be distributed throughout the office, or indeed an entire organisation. The communications required to support this activity have been slow to develop due to a number of reasons:

— product manufacturers wishing to protect their existing user base and hence not producing links to other vendors' equipment;

— lack of internationally developed and agreed interface solutions;

— lack of co-ordination of the user requirement for such developments.

Individual vendors have specified product architectures that allow the interconnection of their own equipment, but connection between different vendors' equipment has always been a problem because of the lack of an agreed interface. What is required is an internationally standardised interface solution that will allow differing technologies to communicate using hardware purchased from different suppliers.

CCITT X.400 MESSAGE HANDLING SYSTEMS

This ultimate goal is now not some distant oasis on the horizon: the standardised interface solution has arrived under the guise of the International Standard Organisation (ISO) work on Open Systems Interconnection (OSI). The OSI framework and its internal operations define a standard interface for the reliable transfer of data between computer systems. Architecturally this work is based around a seven-layer model which effectively separates the functions into logical groups (see Appendix 4 for further explanation).

Within the OSI framework the actual user applications, that is, the functions with which the user will interact, reside at the uppermost layer. X.400 message handling, as defined in the CCITT's X.400 recommendations, is among the first OSI-based user applications to have been defined so far. X.400 defines a method for carrying out medium-independent electronic messaging on a store-and-forward basis between computer-based systems. The term 'store-and-forward' should be noted, as it means that these systems are only suitable for applications that are not real time, ie those that do not require an instantaneous question-and-answer type working. With that proviso, however, these systems can act as a medium for transporting any form of structured or unstructured data. In this way they may act as a platform for many different industry specific applications.

Because X.400 offers a standardised solution, it can also be used to achieve the integration and interconnection of the information transfer technologies which have evolved in today's business environment and because the X.400 recommendations are internationally defined and agreed, it means that companies all over the world will be able to build conforming products secure in the knowledge that they will be able to interwork with other X.400 vendor products. It is likely that very soon – perhaps during the next couple of years – X.400 Message Handling Systems will develop to become the most important part of a worldwide messaging infrastructure. It will also be instrumental in aiding the worldwide adoption of ISO OSI communications standards. This is the only way in which users will achieve the equipment vendor independence for which they all crave.

(Terminology Note: The CCITT X.400 recommendations are a series of eight closely related documents numbered in the range X.400 to X.430. Although the term X.400 is one of the individual recommendations, it is acceptable to use the term X.400 when referring to the series of recommendations as a whole. This convention will be used throughout the remainder of this book.)

PURPOSE AND SCOPE OF THE BOOK

The aim of this book is to give the reader a step-by-step introduction to the topic of Message Handling Systems (MHSs) based on the X.400 recommendations. Chapter 2 looks at the general area of electronic information transfer by introducing the technologies which have been developed and used to date. This base information will give the reader sufficient insight into the subject to be able to compare the potential advantages of the X.400 approach.

Chapter 3 introduces the basic building blocks of an X.400 Message Handling System (MHS). This will establish the required knowledge base for the reader to be able to understand the issues dealt with in later chapters. Appendix 7 specifies more detailed information about certain aspects of the X.400 recommendations that will be of interest to those considering implementing a system. Chapter 3, however, will be adequate for most purposes.

The United Kingdom already has an operative X.400 message handling service – British Telecom's Gold 400. But how will this and other X.400 services integrate to form the UK messaging infrastructure? Chapter 4 looks in detail at Gold 400 and at the existing UK telecommunications legislative structure to assess how this might impact on such developments.

X.400 is now achieving unprecedented interest and implementation, on a worldwide scale. New developments seem to be emerging in all the major-and indeed some not so major – population centres around the globe. It is these very developments which are now instilling people with confidence in the belief that X.400 will succeed and achieve the mass take-up it truly deserves. Chapter 5 strives to give the reader a flavour of some of these developments.

The main theme throughout the book will be to try to address the issues that will have particular bearing on potential end-users. In this context two chapters of particular interest are 6 and 7, which deal with the applications for X.400 systems and the benefits of using X.400, respectively.

However, the implementation and use of X.400 within your organisation might not be as straightforward as some vendors would have you believe. Chapter 8 addresses this topic by considering the stages which should be encompassed by a practical implementation: it also outlines some of the legal issues that have to be considered when utilising electronic messaging for trading transactions.

Following on logically from the previous chapter, Chapter 9 details two implementation case studies from organisations indicating how they have or plan to implement X.400 systems to meet their communication requirements.

Since the development of the original X.400 recommendations (completed in 1984) the CCITT has continued to work on this topic. Their work has now culminated in the publication of a new series of X.400 recommendations which include many new concepts. Chapter 10 looks at the changes and additions to the original X.400 recommendations and attempts to place this new development in context.

Although the theme throughout the book will be X.400, associated areas such as Open Systems Interconnection (OSI), the standardisation process and the groups concerned with standards, will also be considered in the appendices. In this way the book should act as a useful source of background information for anybody new to the general topic of standardised computer communications.

The approach taken at all stages will be to start at the basics and then to build on this knowledge base. More experienced readers therefore, may prefer to overlook some of this background material. However, the main part of the book consists of new and potentially important material concerning the current position, future implementation, use and development of X.400 systems.

2 Electronic Text Information Transfer
The Situation to Date

Systems based on the X.400 recommendations are capable of carrying all types of data, whether it be text, graphic or pure data. Initially, however, it is likely that their predominant usage will be for text transfer, with further exploration of the potential of such systems in the future when experience has been accumulated. In this light it is obvious that X.400 systems will be judged, initially at least, against existing electronic text transfer systems/services in order to ascertain whether they provide a worthwhile option.

Electronic means for transferring text information have existed for some time now, the earliest of course being telex, which appeared in 1932. Through the use of such systems organisations have enjoyed the benefits of improved speed and flexibility in their communications. These improvements are especially apparent when compared with the outmoded postal services.

In order that systems based on the X.400 recommendations can be assessed with a true perspective, it is important to look at the existing options for electronic text information transfer. With this information functions, facilities and potential of X.400 systems can be assessed more clearly. Under this generic title of electronic text information transfer, systems and services such as the following can be considered:

— telex;

— teletex;

— public electronic mail;

— private electronic mail;

— facsimile.

These existing options will be examined in more detail within the following sections.

TELEX

Telex is a service that all organisations and individuals will be readily familiar with. It offers its users a service for the interchange of simple text-based messages on a point-to-point basis. It is essentially a slow speed service which lacks functions and facilities when compared to some recently developed forms of electronic messaging. Recent improvements in the design of terminals, however, have meant that the service has maintained, and indeed in some cases increased, its popularity. Many organisations still rely on telex as their main means of communication in the business environment and hence it has become crucially important for their livelihood.

So why is it so popular? There are really two reasons; first, it is an international service which has a worldwide user base of around 1.5 million spread over 200 countries. Second, it is still the only telecommunications service that provides a hard copy, at both ends of the link, which is internationally recognised as a legal document. This legal status would seem, however, to be without foundation as recently British Telecom stated that they were not aware of a precedent, on which English law is largely based, being set which establishes a telex message as a legally binding document.

Most of the more recent developments in text communications, such as electronic mail (public and private) and teletex, offer a gateway through to telex (though in the case of teletex this will not be available after mid-1988, along with other Interstream services) because of its large installed user base. Access to this existing base is an essential factor in providing the critical mass of users required to get a new service off the ground.

Some of the traditional drawbacks to telex have been addressed by the arrival of computer-based systems. By means of a personal computer or mini/mainframe computer workstation, a telex can be received and read, or written and transmitted without the user having to leave their desk. Moreover, many computer-based telex packages take care of sending and receiving messages in background memory so that the user may carry on with their current task without having to wait for, or to attend to, the receipt of a message.

The advantages of telex are:

— easy to use;

— cheap service;

— large, worldwide user base;

— interface capability with other networks;

— point-to-point communication.

The disadvantages are:

— message preparation facilities can be poor;

— prone to error;

— limited character set;

— slow speed;

— location dependent;

— resource dependent.

Its primary application is in the delivery of short text messages.

TELETEX

Teletex is a service which adheres to an internationally agreed standard for the electronic transfer of documents. It provides for the high-speed transmission and receipt of text-based messages on a point-to-point basis. It can be viewed as a sort of super telex service offering its users improvements in terms of transmission speed, accuracy of messages, enhanced message preparation facilities and better quality hard copies.

The teletex service enables users to prepare text documents on teletex compatible devices such as electronic typewriters or word processors, and then transmit them via a public network to other, maybe dissimilar, devices having teletex compatibility or via a gateway to the national and international telex services. Local mode operation (text preparation, editing, printing etc) is not normally affected by the transmission or receipt of a document, as this is a memory-to-memory communication system.

The teletex service was conceived by the German Bundespost in the mid-1970s and developed in co-operation with Siemens. The development was due to two major problems which faced Bundespost.

These were:

— linking of dissimilar text systems, with the associated problems of incompatible character sets, protocols and formatting controls;

— the outdated telex service. Telex is slow, prone to error, has poor presentation with equipment which lacks a good user interface and text creation facilities. This latter problem has been eased slightly with modern telex terminals.

It was from these beginnings that teletex evolved and was subsequently taken on board by the CCITT as an international standard for text communication.

For users who have word processing facilities and require access to this service there are a number of options, depending on the type of system that is available.

— *teletex terminal*. This does not have to be a dedicated terminal; it can range from a simple electronic memory terminal to a word processor, business computer or combined workstation.

— *adaptor or 'black box'*. This can add teletex capability to an existing word processor (much in the same way as a telex adaptor), sophisticated telex terminal or computer by means of a software package.

— *multi-user*. Where more than one user on the system needs access to teletex some means of distributing the facility must be found. This is usually via a cluster controller, local area network (LAN), or PABX.

Although teletex was initially devised as a replacement for telex, this has not occurred in any country except Germany (in excess of 17,000 terminals). British Telecom's launch of the UK teletex service in 1985 has failed to capture user interest (fewer than 300 registered users in the UK) and now they are to remove the Interstream telex gateway facility (mid-1988).

The advantages of teletex are:

— internationally standardised service;

— in-built error detection;

— high-speed service;

— good message preparation facilities;

— messages received when equipment is unattended/'no power' receipt;

— extended character set;

— document not message based.

The disadvantages are:

— small user base in the UK;

— can be expensive to get started;

— document not message based;

— location dependent.

Its primary application is for the fast delivery of text-based documents.

PUBLIC ELECTRONIC MAIL

An electronic mail (E-mail) service will provide its users with person-to-person communication service for text-based messages. When you register with an E-mail service you will be allocated a password-protected mailbox or message storage area on the system's host computer. This means that if you are unavailable any messages will be stored until you can access them at a later time or date. Only people who know the password for a particular mailbox can gain entry, so the systems are quite secure. E-mail services also provide their users with access to on-line databases, noticeboard facilities, the telex service and other useful features.

These services, such as Mercury Link 7500 (was Easylink), Quik-Comm (GE IS), Comet (Istel) and One-To-One, have been allowed to evolve in the UK since the 1981 Telecommunications Act (see Chapter 4 for a full introduction to the UK regulatory framework). This enables organisations to provide services over the public switched telephone network (PSTN) and the packet switched service (PSS), whether or not they compete with the services which British Telecom (BT) offer, in this case Telecom Gold.

To gain access to these systems requires that the user is registered with the service and has a computer with the appropriate software package. The user may utilise a micro or be connected to a multi-user mini/ mainframe installation. A communications software package is required for the micro system in order to allow it to download messages, files and information from the E-mail service computer and uploading text prepared off-line; the package may be tailored to the particular service or be able to be used in any environment. An approved modem, acoustic coupler, Dataline connection to the local PSS packet switching exchange (PSE) or a multistream service can be used for dial-up access to the E-mail service.

Although such services offer reasonable features and facilities, they seem to fail to meet the user criterion in the area of interconnection. Such systems and their associated user bases have remained unable to intercommunicate between themselves since their introduction. This situation seems unreasonable as computer systems generally seem to be moving towards higher levels of integration.

The advantages of public electronic mail systems are:

— low starting costs, especially if a micro is available;

— access to other value-added services, such as telex, databases, etc;

— contacting people who spend time out of the office, transferring text via portable PCs and acoustic couplers;

— the message security offered by the password-protected mailboxes;

— reasonably large user bases.

The disadvantages are:

— high running costs;

— no connection between different vendors' systems.

Public electronic mail services are mainly used for fast, location independent delivery of messages rather than extensive documents.

PRIVATE ELECTRONIC MAIL

The provision of electronic mail facilities within private organisations is now an established means of achieving fast and efficient intra-business communications. Such private electronic mail systems offer either similar or more advanced features than those of the public electronic mail services, but are provided on corporate hardware. In the past, organisations have been forced into the purchase of incompatible systems and this has led to special difficulties when interconnection is required. The task of integrating a number of corporate electronic mail systems – from different sources – is not easy. In such cases it is difficult to achieve acceptable results and can often lead to considerable expense, due to the one-off nature of such problems.

For an organisation wishing to set up its own internal electronic mail service its options fall into one of two categories. These are:

— electronic mail facilities provided as part of an integrated office system such as DEC's All-In-One or Data General's Comprehensive Electronic Office (CEO);

— licensed electronic mail software from an electronic mail vendor: an example here would be Istel's COMET which is available in two forms, one to run on DEC systems and another version for IBM systems. Such packages may be customised to cope with individual customer requirements: an example here might be the integration of the mail system with existing word processing facilities.

Integrated office systems, as their names suggests, attempt to integrate all the logical office facilities within one system, such as:

— electronic mail;

— word processing;

— graphics;

— spreadsheet;

— communication to outside services such as telex, Prestel, etc;

— document filing and retrieval.

Such systems are extremely powerful within their own environment and will allow the integration of text, data and graphics within documents which can then be transferred between users. The advantages of private electronic mail are:

— minimal running costs;

— security, reliability and availability are high;

— possible to access other value added services;

— location independent;

— can be message or document based.

The disadvantages are:

— high initial costs;

— in-house support required;

— lack of interconnection with external systems and those from other vendors;

— lack of impact.

Private electronic mail systems are mainly used for intra-organisation document or message communications.

FACSIMILE (FAX)

Facsimile is a service which can transmit and receive prepared text and graphic information. The process employed is straightforward; in simple terms the document image is encoded as a series of black or white elements. A light-sensitive scanner reads the original document, dividing the image into thousands of squares. This scanned image data is then converted into digital impulses which are in turn converted into analogue signals, via modem, suitable for transmission over the public telephone network (national or international). The receiving machine will interpret the signal and recreate the original image.

In the early days facsimile machines were slow and generally manufacturers produced their own incompatible implementations. With the recent rapid developments in electronics hardware the speed of the machines has been significantly improved, but the problem of incompatibility had to be tackled by the adoption of international standards. The CCITT (International Consultative Committee for Telephone and Telegraph) some 15 ago developed the first international standard, Group 1, and since then the market for facsimile has grown rapidly.

All facsimile machines are now manufactured according to the design standards agreed by the CCITT, and terminals have been divided into four groups.

Group 1, established in 1968, used analogue transmission techniques and took nearly 6 minutes to send an A4 page. These machines are now virtually obsolete.

Group 2, established in 1976, also uses analogue transmission techniques but takes 3 minutes to send an A4 page and has a better image quality at the receiving end. These machines too are rapidly being replaced by, or passed over for, Group 3 machines.

Group 3, established in 1981, is now the worldwide machine standard, especially for international communications. By utilising data compression techniques these machines make the best possible use of the transmission medium, hence fast document transmission times. A4 document transmission time has been reduced to less than a minute with the emergence of Group 3.

Group 4, The CCITT set these standards in 1984 with the purpose of allowing facsimile transmission to take account of the developing public digital telephone networks. When a fully interconnected digital network has been established, the use of Group 4 machines will become viable and should allow very rapid transmission of documents. Transmission times for an A4 document have been quoted as low as 5 seconds via Group 4 technology.

The emergence of these standards has had a massive impact on the user adoption of facsimile transmission. In only a very short period its use has escalated to the point were it now rivals telex as the most popular means for document transmission.

In the past the only method of sending a facsimile was to produce or obtain the required text and/or graphics in a hard copy form and then to go to the facsimile machine and send the document. This generally involves a person having to leave their desk or place of work and then having to spend time establishing the phone connection before being able to send the information. This problem is now going to be addressed by the recent arrival of personal computer PC-to-fax links.

Because of the rapid proliferation of facsimile terminals and personal computers into the office environment it was only a matter of time before someone devised a method of exploiting the attributes of both devices. A PC-to-fax link consists of a software package and a modem for the PC which allows you to digitise the information on the screen into the required format for a fax transmission. The modem is then used to send the information via the PSTN to a dedicated fax terminal or another PC with a fax link. The software can handle the task of sending the information as a background task and so the sender is free to continue with another task. Hence with a PC, the required software, a modem and a phone link a person will be able to send messages without ever having to leave their desk.

The advantages of facsimile transmission are:

— can transmit both image and text information separately and together in the same document;

— easy to use;

— no special training for users;

— high speed transmission becoming commonplace;

— low transmission costs.

The disadvantages are:

— not secure;

— prone to errors;

— does not integrate into existing systems, PC-to-fax links will help;

— location dependent;

— resource dependent.

Its primary application is in the fast delivery of mixed text and graphics documents of varying length.

ARE THE CURRENT FORMS OF ELECTRONIC INFORMATION TRANSFER SUCCESS-FUL?

It has already been stated that within business and commerce there is a need to accomplish the efficient and accurate transfer of information between locations. Any improvements wrought in the flow of information will generally produce gains in terms of productivity and the levels of success achieved by organisations. In this context, the use of electronic information transfer within the business environment has been very successful, with users reaping many benefits in terms of improved communications and productivity.

It is interesting to look at some of the figures from the recent National Computing Centre's Office Automation (JOA) study that indicate benefits expected and achieved from OA projects involving technologies such as telex and internal/external electronic mail. One point that did seem to emerge from this project was difficulty users have had in trying to quantify the impact of some of the technologies which have been discussed so far. Because such features are not directly replacing an existing office function, but rather yield extra functionality, there is no existing standard by which to judge them. Taking the example of word processing facilities replacing those of the typing pool, it is easy to see that comparative timings can be made; and views can be canvassed as to the ease of use of the new equipment etc. In most cases electronic messaging is deemed to have significant benefits, but they are difficult to assess accurately. An example here might be where sending an electronic mail message can overcome the inconvenience of repeated telephone calls to someone who is very busy. With this difficulty in mind however, the results obtained from the study are still significant. Figure 2.1, shows the objectives or expected benefits which the users anticipated would be achieved from specific types of OA projects.

In all cases the biggest expected benefit is in productivity gains, ie improvements wrought in working practices by the use of this type of function. It is also apparent that with projects involving electronic messaging, whether telex, teletex or internal/external mail systems/services, improved communications is a major factor which is expected to be achieved. This final point is particularly apparent with projects involving internal mail; this is to be expected because experience has shown that some of the worst aspects of business communications are those internal to organisations. This is somewhat surprising, as this area can be crucial to the performance of a business, particularly a large one that is formed of many autonomous departments.

Moving on to look at the overall satisfaction ratings by project type (see Table 2.1) we can see that projects involving the use of electronic messaging are generally deemed to be quite successful. In fact almost 80 per cent felt that projects involving these technologies either exceeded or fulfilled expectations. From these figures it can be reasonably stated that existing forms of electronic messaging are generally quite successful in improving communications both at the inter- and intra-organisational level.

It is also interesting to look at Table 2.2, which defines the future expansion and provision of OA facilities. Almost 50 per cent of the groups questioned were going to provide or expand their electronic mail facilities, while other forms of electronic information transfer (such as telex, facsimile, etc) figure less prominently. This would seem to indicate that organisations are looking towards computer-based messaging facilities as their future basis for business communications.

All the figures mentioned so far seem to indicate general satisfaction with electronic mail/messaging facilities and that they are likely to play a significant part in the future development of OA systems. So why look towards X.400 based message handling systems? To establish why this approach should be taken we must look at some of the problems associated with existing systems.

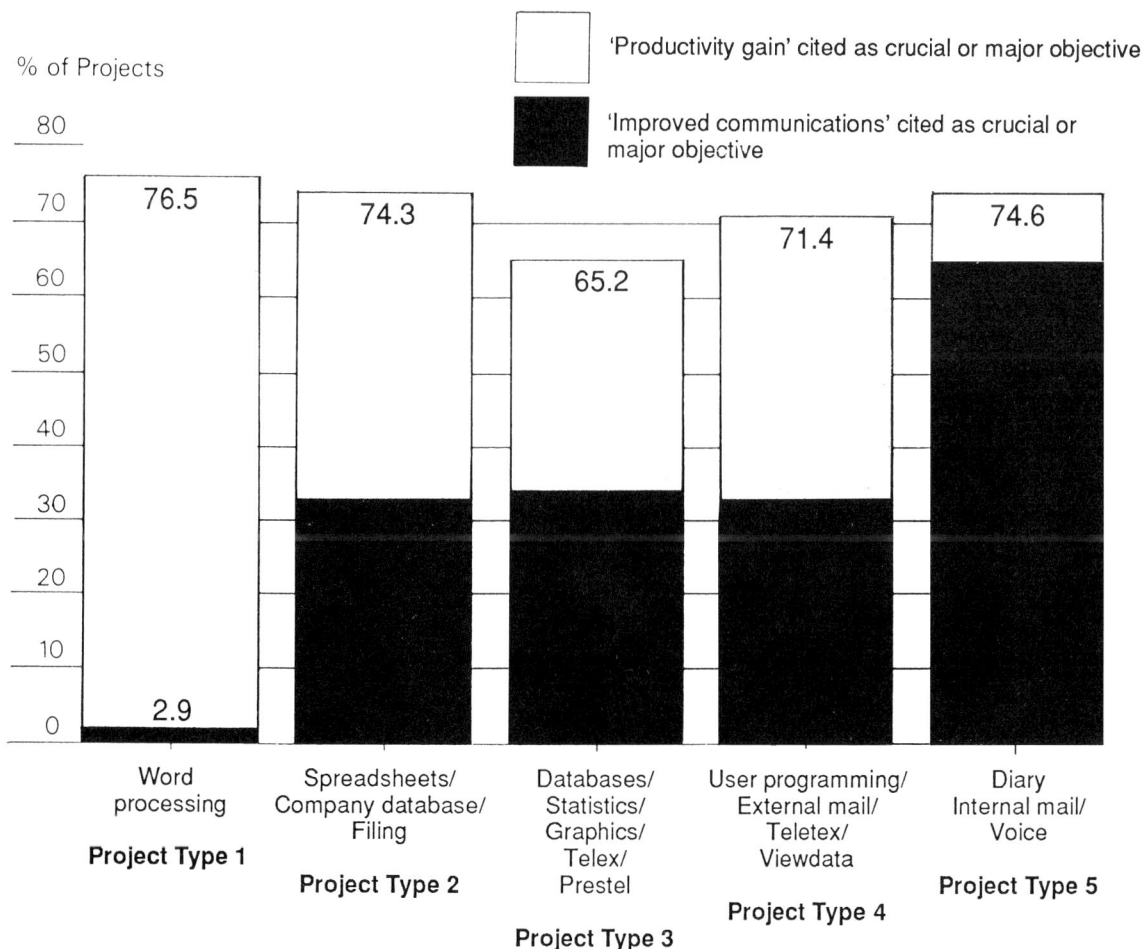

Figure 2.1   Objectives or Expected Benefits from OA Projects by Project Type

	Exceeded expectations	Fulfilled expectations	Fell short of expectations	Base
	%	%	%	
Project Type 1	14.8	77.8	7.4	27
Project Type 2	6.9	79.3	13.8	29
Project Type 3	5.3	78.9	15.8	19
Project Type 4	20.0	60.0	20.0	15
Project Type 5	20.0	53.3	26.7	45
TOTAL	14.1	68.1	17.8	135

Table 2.1   Overall Satisfaction Ratings by Project Type

	%	Base
Word processing	56.6	189
Database	50.3	
Electronic mail	47.1	
Spreadsheets	40.2	
Integrated office software	21.2	
Electronic filing /retrieval	8.5	
End-user computing	7.4	
Telex	7.4	
Graphics	6.9	
Viewdata	6.3	
Diary systems	5.8	
Micro to mainframe communications	5.3	
Voice messaging	3.7	
Facsimile	3.2	
Communication links to external services	3.2	
Data transfer	2.6	
Specialist packages	2.6	
Statistical packages	2.1	
CAD/CAM	1.6	
Data capture/collection	1.1	
Modelling	1.1	
Cellular radio	0.5	
Text handling	0.5	
Information engineering	0.5	
Knowledge engineering	0.5	
Video conferencing	0.5	
Electronic typesetting	0.5	
On-line systems	0.5	

Table 2.2   OA Functions to be Provided or Expanded Within Three Years

WHAT ARE THE PROBLEMS WITH TODAY'S ELECTRONIC MESSAGING SYSTEMS?

In each of the sections on the individual electronic messaging technologies a summary of the advantages and disadvantages was provided. Reference to these will show that invariably each one offers a reasonable service to the user and that the main disadvantages quoted surround the areas of cost, lack of facilities, small user base, interconnection and integration.

The cost of services/systems has always been a source of contention between suppliers and users. The cost of computer-based hardware has reduced drastically with the emergence of LSI and VLSI (LSI = large scale integration; VLSI = very large scale integration, and this in turn yields scope for improving the facilities offered by such systems. Because of this, the cost of hardware has appeared to remain at the same level while in real terms this represents a reduction because newer products will offer significantly increased performance for the same cost. The costs of computer-based messaging services may appear high but increasingly suppliers seem to be providing additional functionality, or – to use the popular term – further 'added value'. It is this added value (eg access to external databases, noticeboards, etc) that can sometimes be crucial in justifying the expenditure on this type of service. Although the points mentioned here would seem to offer some mitigation for the first two disadvantages, quoted at the beginning of this section, the latter two have yet to be successfully addressed by any system or service.

The lack of interconnection abilities and/or failure to integrate with other OA equipment is something which has yet to be addressed by all means of electronic information transfer except telex. Telex, however, although it seems to integrate with most systems, except facsimile, has its own problems because it lacks facilities/functionality and speed when compared to computer-based systems. The problems of lack of interconnection, between public electronic mail services for example, mean that they are instantly restricted in terms of the potential user base which can be addressed. In the UK there are several services available, none of which will talk to each other. The integration of differing technologies is the only way in which their full individual potential, improving communications and productivity, can be realised. Facsimile is particularly guilty in this area, mainly because it is based upon the transfer of scanned rather than the more common text-based information.

In the Price Waterhouse *Information Technology Review 1987/88* an indication of the communications problems facing UK computer installations was provided (see Figure 2.2). For the purpose of this argument only those problems with particular reference to the equipment or technologies will be relevant. Other organisational problems will be considered later, specifically with reference to X.400 systems. The categories of interest here are: limitations imposed by cost and practicality (ie the cost and practicality of achieving such a link); integration of differing technologies; knowing what equipment to buy; and knowing when to buy it (assessing obsolescence). All of these will be common problems to organisations operating some form of electronic messaging, which they have found does not fulfil certain criteria because of a lack of integration with other equipment or a requirement for interconnection. The occurrence of these problems indicated in Figure 2.2 is certainly significant in the range of companies interviewed.

WHAT DOES THE X.400 SOLUTION OFFER?

The CCITT's X.400 Message Handling System (MHS) recommendations define the generic architecture and underlying protocols for an all-purpose message handling service. These recommendations are internationally agreed and therefore provide a common basis from which the developers of office systems and electronic mail services can produce systems which will be able to be interconnected and so interwork.

The system defined in these recommendations is basically a store-and-forward electronic messaging service with the ability to transfer any type of information, whether it be text, image or pure data. The 'store-and-forward' term is significant; it means that such systems are not real time, because the time when a message is offered to such a service for transport bears no fixed relationship to when the message is delivered to the intended recipient or system. X.400 systems are effectively separated between the user applications and the underlying means of message transfer. This feature allows multiple-user applications to be run over the same link. Initially at least the major application will be interpersonal messaging; this will operate in a similar manner to existing public electronic mail services, allowing the interchange of messages and documents between human users.

Because these recommendations are internationally agreed and available in the public domain, it means that for users of electronic messaging X.400 should be able to act as the basis upon which a worldwide

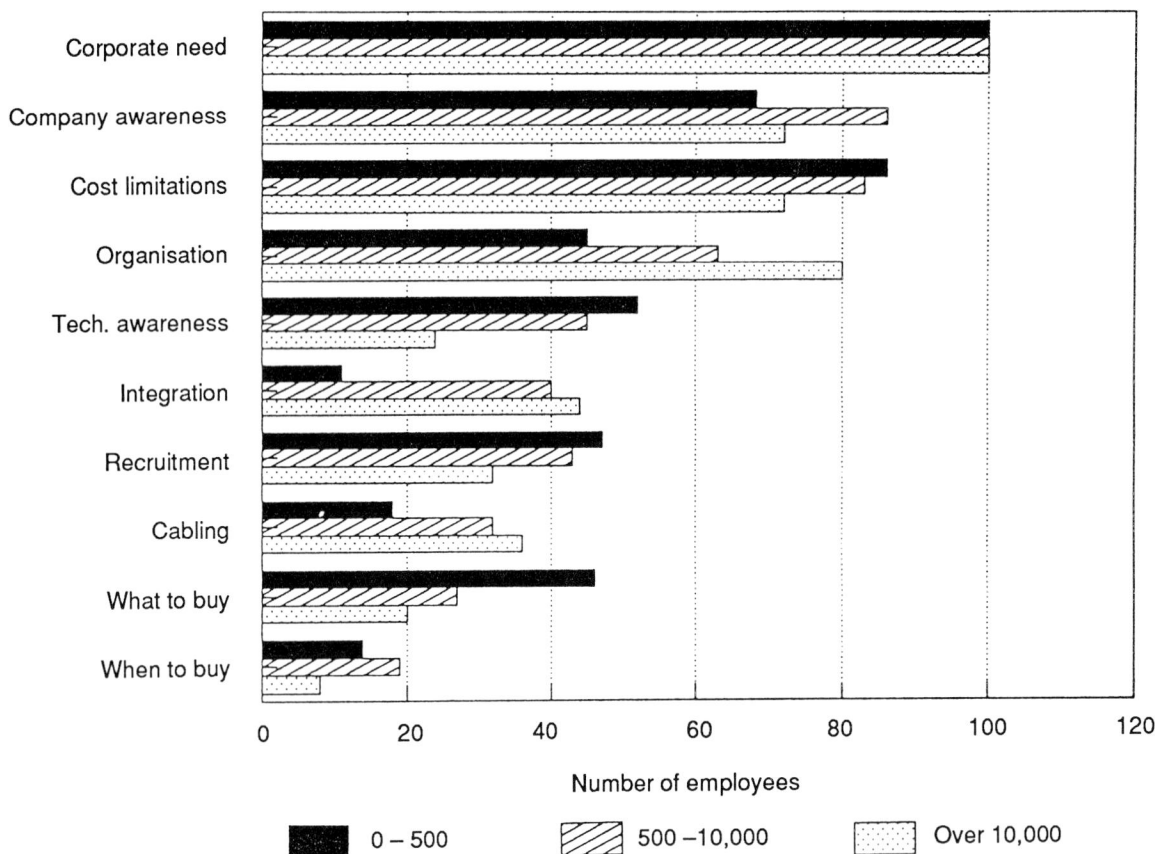

Figure 2.2   The Top Ten Problems Facing UK Computer Installations Implementing Communications Systems

messaging infrastructure will be established. Existing messaging systems and services can be interconnected via the use of X.400 protocols, which in practice will mean:

— service providers or PTT provision of connections to other X.400 systems around the globe with access units from their X.400 based services to telex, facsimile and teletex;

— equipment vendors will either provide X.400 gateways or complete systems based on X.400 which will provide for the interconnection of a private system with another or with a public message handling service.

The X.400 solution is future proof because any development of the current recommendations will maintain compatibility, and companies will no longer be confronted with the dilemma of what to buy and when to buy it. Users will be able to enjoy the vendor independence for which they all crave. Also, because vendor products will conform to the same standard, they can be selected purely on merit rather than on their ability to interconnect or integrate with existing product purchase decisions.

It was stated earlier that X.400 systems can act as a platform for a wide range of applications, other than just electronic mail. This really is the key to the usefulness of such systems because applications can be updated, replaced or added as is deemed necessary. In this way an X.400 system can grow in the future to encompass applications which have not yet even been considered. This means that when considering such a system it can be looked upon as an all-purpose data transfer mechanism with enormous potential for future development.

It should by now be apparent that the adoption of systems based on X.400 has some substantial benefits when compared to existing means of electronic information transfer, such as:

— internationally standardised solution;

—   future proof;

—   a basis for industry specific applications, not just electronic mail;

—   interconnection and integration of existing technologies.

The argument in favour of the X.400 solution is really only being touched upon in this section, but will be extended greatly throughout the remaining chapters of the book.

3 X.400 – A Brief Technical Guide

The purpose of this chapter is to give the reader a basic understanding of the concepts and terminology of the CCITT X.400 recommendations. Armed with this knowledge the explanation of the applications and benefits of X.400 will be more easily achieved.

The level of understanding required about this topic will vary immensly within organisations. Some people will only require a grasp of the basic concepts, while others with a responsibility for product/ equipment purchases will need a deeper insight into the subject. Because of this problem, it was decided to introduce the technical aspects of the X.400 recommendations in two stages. The first part must be viewed as essential reading, while the latter (see Appendix 7) will depend on the level of the reader's interest in the subject.

This initial stage will establish the basic concepts required to comprehend the issue being dealt with. The second technical section, Appendix 7, will deal with the subject in greater depth by looking in detail at the architecture of the model and the possible implementation options. This information will also be of use during any purchasing decisions.

DEVELOPMENTS IN MESSAGE HANDLING SYSTEMS AND STANDARDISATION

The development of computer-based Message Handling Systems (MHSs) began in the mid-1960s with a number of groups developing systems to allow the exchange of electronic messages within computer networks. These developments resulted in product manufacturers producing such mail systems for existing hardware. This is a perfectly acceptable type of solution provided that the systems to be interconnected are produced by the same manufacturer.

The development of standardised solutions for electronic messaging systems began in the early 1980s with the International Federation for Information Processing (IFIP), who developed a basic model of a store-and-forward message transfer system. These concepts were rapidly adopted and developed by the three major standards groups, ECMA, CCITT and ISO (see Appendix 3). The development work of these respective organisations differed at detail levels because of the service spectrums they were expected to address, eg CCITT – administrations and service providers, ISO – the private messaging environment. All these messaging systems were based upon the principles of Open Systems Interconnection (OSI – see Appendix 3) with its seven layer architecture model and protocol specifications. Message handling, being a user application, resides at the seventh layer, ie the highest layer, of this model. After a certain amount of co-ordination between the groups a situation evolved whereby the CCITT MHS recommendations are complementary to the ECMA – MIDA (MIDA = Message Interchange Distributed Application) and both are incorporated into the ISO MOTIS standard (MOTIS = Message Oriented Interchange System).

The CCITT have a traditional responsibility for the standardisation issues surrounding international telecommunications. Their work is carried out in four-year study periods at the end of which recommendations are published in the form of books. The colour of these publications indicates the study period from which the recommendation emanates. For example, X.400 was produced during the 1981—1984 period and is a book with a red cover.

When, in 1984, the CCITT ratified their set of eight MHS recommendations (numbered non-sequentially X.400 to X.430) it was decided that the ISO MOTIS work should adopt the same words and principles wherever possible. MOTIS, however, could be considered as a compatible superset of X.400, extending it into the private messaging domain. This is specifically indicated by the ISO provision for links between private messaging domains whereas CCITT recommendations do not define these (full explanation later in this chapter).

CCITT were quick to realise that the 1984 version of X.400 MHS contained a number of inaccuracies and ambiguities. Although most of these have been highlighted in issues of the X.400 *Implementor's Guide* (special guidelines highlighting problems and implementation choices) it was still the intention of CCITT to have as near perfect as possible a set of documents. To this end the CCITT continued to develop and extend the scope of their existing recommendations, as ratified in 1984, during the subsequent 1985—1988 study period.

Steady development of MOTIS continued after 1984 until 1986, by which time it had achieved DIS (draft international standard) status. At this point ISO decided to freeze all work on the existing documents in order to work with the CCITT during 1987 to develop joint recommendations/standards for messaging. The new work included input from both the CCITT and ISO in order to achieve a globally acceptable generic standard. Early in the development of these new 1988 X.400 recommendations it was stipulated that backward compatibility with 1984 should be ensured so as not to compromise investment in existing systems.

Implementations based on these new recommendations are some years away at the time of writing (early 1988). Indeed, looking at the number of implementations of the existing 1984 recommendations it is difficult to see 88 MHS having a substantial impact before 1991. For these reasons the new recommendations are only considered in passing while the rest of this chapter—and indeed this book—is based on the issues surrounding the existing 84 X.400 recommendations.

(Note: For further information about 88 X.400 refer to Chapter 10.)

THE BASIC CONCEPTS AND TERMINOLOGY

As with all International Standards and recommendations, X.400 is riddled with jargon likely to confuse the casual observer. This guide will attempt to avoid this approach by defining all terminology and drawing on clear, practical examples where possible.

The X.400 Recommendations

The CCITT X.400 series of Message Handling System (MHS) recommendations is comprised of eight closely related documents which define a generic architecture for medium-independent MHSs.

In order, the eight documents are:

X.400 — specifies the network model comprising:

- user agents;

- message transfer agents;

- message transfer;

- interpersonal messaging services.

It also details the protocol layering of an MHS.

X.401 — describes the basic service elements and optional user facilities of the MHS. This recommendation gives an overview of the Interpersonal Messaging (IPM) and Message Transfer (MT) services which provide the above functions;

X.408 — specifies the rules for converting one information type to another, eg telex character codes to IA5 (ASCII). This recommendation was largely left for further study in 1984, but will be remedied in the 88 X.400 recommendations;

X.409 — defines the presentation transfer syntax rules used by application layer protocols in an MHS. These rules specify; a notation which is used for describing the MHS protocols; and a set of encoding rules which are applied to the description to produce a string of bits suitable for transmission. This notation and the encoding rules are also used in other OSI applications—they are known as ASN.1 (Abstract Syntax Notation-1);

X.410 — details the facilities and protocols used by Remote Operations and the Reliable Transfer Server which make up the lowest layers of the sublayered MHS architecture;

X.411 — specifies both the service and protocol for the Message Transfer Layer (MTL) within the MHS model;

X.420 — specifies the content protocol for interpersonal messaging services and includes the specification for a simple formatted document description (SPD);

X.430 — details the specialised means of access for teletex terminals which allows them access to the functions of the MHS.

For a newcomer to the subject, the X.400 recommendation itself serves as a suitable introductory document to message handling. It defines the main components of a system and their basic interaction without descending too deeply into the complex subject of protocols.

WHAT CONSTITUTES A MESSAGE HANDLING SYSTEM?

Message handling systems may be viewed as store-and-forward electronic mail systems, but these are not real time systems and therefore not suitable for real time applications. X.400 systems will allow people or computer application processes to communicate by submitting and receiving messages which can consist of any structured or unstructured data. Because the X.400 Message Handling System (MHS) recommendations have been internationally agreed they can provide a standardised electronic mail system for the integration and interconnection of all today's disparate office systems and equipment.

Introducing the Jargon

As was stated earlier, X.400, in common with all International Standards and recommendations, is full of jargon which requires definition before it is possible to grasp some of the basic concepts.

Consider the individual components of an X.400 MHS as indicated in Figure 3.1:

MESSAGE TRANSFER AGENTS (MTA)

These are store-and-forward message exchanges whose functions are analogous to those of sorting offices within the public postal service. They are interconnected and are known collectively as the Message Transfer System (MTS).

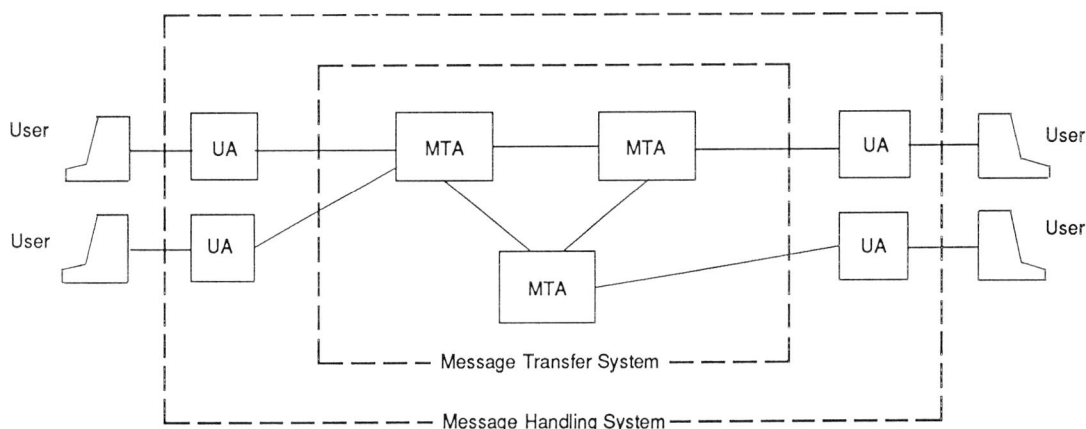

Figure 3.1   The Basic Components of an MHS

The interconnection of MTAs is via any media capable of supporting a full OSI network protocol.

USER AGENTS (UA)

These provide the user interface to the MTS and may include local editing, temporary storage and archiving facilities. The UA accepts a message from the user and then employs the MTS to route the message to the recipient.

USER

This is, in X.400 terms, not a human user; rather it is the equipment or service which is using the MHS to achieve message transfer. It may be a dumb terminal, local host or a third-party electronic mail facility connected through a public data network. The interconnection of UAs and MTAs is called a Message Handling System (MHS).

It must be emphasised that the X.400 recommendations do not specify the manner in which the components of an MHS are to be implemented. The MTAs and UAs are merely groups of functionality placed together because they appertain to a particular type of function, ie the UA deals with user functions while the MTA is concerned mainly with the message transfer function.

The Message Handling System Model

Conceptually, the store-and-forward message handling model has a great similarity to the postal service, which employs a form of sorting office and delivery agents. Consider the cycle and the elements indicated in Figure 3.2, which details the transfer of a letter via the postal service.

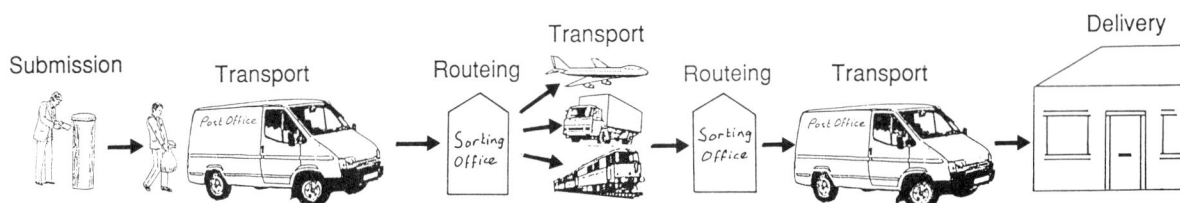

Figure 3.2   The Basic Model of the Postal Service

A person will deliver a letter or message to a post box, and the letter will be addressed with sufficient information to achieve delivery to the intended recipient. The letter is then transported or submitted to a sorting office. Here a route for the letter will be established from the envelope address. The method of transport over this route will be dependent on the class of service specified (eg 1st or 2nd) by the type of stamp affixed to the envelope. The letter is then transported to a sorting office in the recipient's area where the final sorting process is accomplished and delivery to the recipient's letterbox achieved.

With this example in mind it is possible to see the analogies of this service with the Message Handling System model, as shown in Figure 3.3.

It is obvious that the process of message transmission in an X.400 system breaks down quite neatly into three distinct phases. These are now dealt with below:

Message Submission

The user, through interaction with the 'originating UA', will produce a message for despatch. This part of the operation is analogous to a person producing a letter and depositing it in their post box ready for collection. The UA will then 'submit' the message to its MTA in an envelope which contains all the information required by the MTS to route the message to its intended recipient.

Message Relay

The MTA will analyse the information on the envelope in order to determine the route across the MTS. MTAs can be considered as being analogous to sorting offices in the postal service because they look at

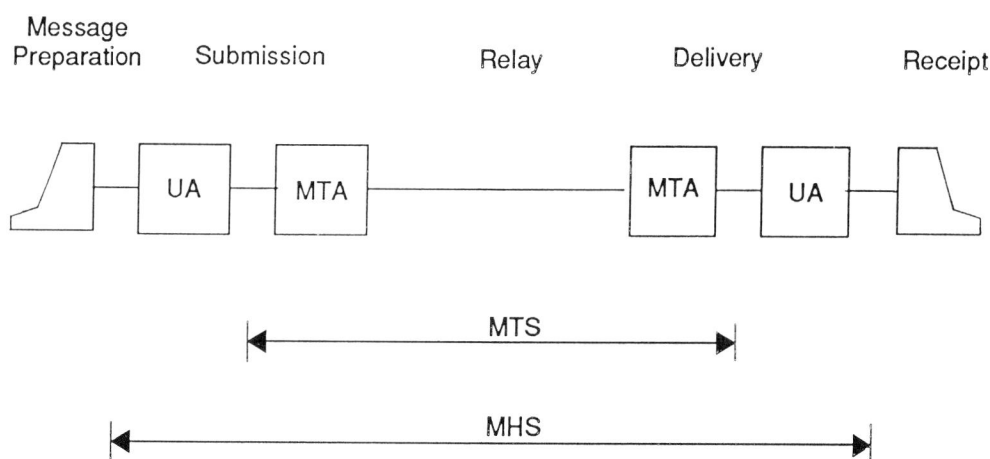

Figure 3.3   The Message Handling System Model

the envelope information in order to determine the intended recipient. The actual process of message relay may involve one or more MTAs. The messager will time and date stamped so that a record of their route across the MTS is maintained. In a similar way to specifying a class of service for letters (eg 1st or 2nd class stamp) it will be possible to specify the urgency of delivery in a practical X.400 system, eg ranging between urgent and non-urgent.

Message Delivery

The message will eventually progress through the system until it reaches MTA associated with the intended recipient. When this is achieved the MTA will 'deliver' the message to the recipient's UA in a similar manner to the way in which the postman will deliver a letter to the recipient's letterbox. The final stage of the transfer will be complete when the message is received, which is achieved when the user accesses the message in order to read it.

Introducing the MHS model with this simple analogy is a reasonable way of conveying the basic concepts. The terms to remember here are 'submission' and 'delivery':

— Submission of a message takes place when the originating UA transfers a message plus its submission envelope to its MTA. The submission envelope contains the information that the MTS requires to achieve message delivery.

— Delivery of a message is said to have taken place when an MTA transfers the message to the recipient's UA.

A message is only said to have been received in X.400 terms when the recipient has accessed it in order to read it.

MESSAGE HANDLING SYSTEM SERVICES

In order that X.400 MHSs can be as application-independent as possible the services provided by them are separated into two distinct groups. In this way the functions of the service which appertain to a specific user application are separated from those which provide the underlying message transfer function. The two types of service provided by the MHS are:

— Message Transfer Service or MT service (never shortened to MTS to avoid confusion with Message Transfer System acronym);

— User Services.

Message Transfer (MT) Service

This is the generic service which is operated between MTAs and UAs to provide an application-independent, store-and-forward message transfer service. This service can carry any form of structured or unstructured data as it is only concerned with the envelope information unless it is expressly requested to carry out some form of content conversion. It is this ability to carry any form of data that allows X.400 to act as a basis for applications other than just straightforward text messaging.

The basic MT service enables UAs to access and be accessed by the MTS in order to exchange messages. When a message is to be transferred it is given a unique message identifier so that if the message cannot be delivered the originating UA can be informed.

The MT service consists of a number of service groups which except the Basic group, are optional. These groups are, in turn, subdivided into a series of service elements. Each of these service elements performs a specific function within the MT service. The service groups are formed from service elements which together perform a specific aspect of the MT service. The service groups are:

— *Basic.* These are concerned with the unique identification of the message, the date and time of message delivery and submission and the type of contents.

— *Submission and delivery.* These are concerned with the specification of the type of delivery which is required (eg speed, number of recipients, etc) and whether any delivery notification is applicable.

— *Conversion.* Whether or not message content conversion is required (dealt with in more detail later).

— *Query.* To establish whether it is possible to deliver a message, ie it is suitable to be displayed on the recipient's terminal.

— *Status and inform.* This service group allows messages with certain attributes to be delivered only to specific UAs and gives the UA the capability to indicate to the MTS that it is not ready to accept message delivery.

Although generally the MT service operates purely as a means of transferring messages, there is a special case where it takes a more active role in the formatting of the message in order to achieve delivery. 'Content Conversion' is where the MT service will re-format the content part of a message so that it can be delivered to a non-X.400 terminal. An example here might be the delivery of an X.400 message to a telex machine or a teletex terminal. Access to such services will be provided by the national PTT's X.400 system utilising gateways. The recommendations concerning the conversions to be made between information types, eg IA5 to telex etc, are incomplete, and specified for further study. This is not such a problem, however, as the concepts of such conversions are well understood. The types of conversion to be offered by an MHS, however, have already been finalised. These are shown in Table 3.1.

User Services

Although the MT service is comprehensive and complete there has been only one user service defined so far. This is the Interpersonal Messaging Service (IPM Service) which provides its users with electronic mail/messaging facilities similar to those of existing public electronic mail services.

Each user service will be defined to accomplish a specific user task or application (such as electronic messaging) and for each one defined a new class of UA must also be defined. The design of these new UAs is not hindered by having to be compatible with those of existing user services. They merely have to be able to interface successfully with the underlying message transfer system. The definition of a new standardised User Service will be in response to the identified needs of the user community.

Interpersonal Messaging Service

The Interpersonal Messaging (IPM) service provides its users with electronic mail/messaging facilities. The service is built upon the MT service and is provided by a class of UAs called IPM UAs. The service also allows intercommunication with telex and telematic services, such as facsimile and teletex via gateways or access units. Also, as X.400 links become established with other forms of proprietary and private electronic mail systems, an IPM user will be able to exchange messages with these.

IPM UAs make use of the Basic group of MT facilities while also permitting the optional ones to be requested. In addition to these service elements IPM UAs also provide other capabilities which make up the IPM service. These groups are:

From \ To	TLX	IA5 Text	TTX Basic	TTX Optional[1]	G3 Fax Basic	G3 Fax Optional[1]	TIF0 Basic	TIF0 Non-basic[1]	Videotex	Voice	SFD	TIF1 Basic	TIF1 Non-basic[1]
TLX[5]	–	a	a	a	a	a	a	a	b	b	a	a	a
IA5 Text	b	–	b	b	a	a	a	a	b	b	a	b	b
TTX Basic	b	b	–	a	a[4]	a[4]	a[4]	a[4]	a	b	b	a	a
TTX Optional[1]	b	b	b	b[2,3]	a[4]	a[4]	a[4]	a[4]	a	b	b	a	b[2,3]
G3 Fax Basic	c	c	c	c	–	a	a	a	c[6]	c	c	a	a
G3 Fax Optional[1]	c	c	c	c	b	b[2,3]	b	b	c[6]	c	c	b	b
TIF0 Basic	c	c	c	c	a[4]	a[4]	–	a	c[6]	c	c	a	a
TIF0 Non-basic[1]	c	c	c	c	b	b	b	b[2,3]	c[6]	c	c	b	b[2,3]
Videotex	b	b	b	b	a[7]	a[7]	a[7]	a[7]	FS	FS	FS	FS	FS
Voice	c	c	c	c	c	c	c	c	c	FS	c	c	c
SFD	b	b	a	a	a	a	a	a	b	FS	–	a	a
TIF1 Basic	b	b	b	b	a[4]	a[4]	a	a	b	b	b	–	a
TIF1 Non-basic[1]	b	b	b	b[2,3]	a[4]	a[4]	b	b[2,3]	b	b	b	b	b[2,3]

– No conversion.
a Possible without loss of information.
b Possible with loss of information.
c Impractical.
FS For further study.
1 Specified in the relevant Recommendations.
2 No information is lost if the originating and recipient terminals have the same optional functions.
3 Information may be lost if the originating terminal uses optional functions that the recipient terminal lacks.
4 Information may be lost due to the difference between the printable and reproducible areas.
5 The WHO ARE YOU character is assumed to be a protocol element used for communicating with the telex terminal and not part of the message's content.
6 It may be possible with loss of information, if the recipient terminal has the capability of the photographic type of information.
7 When converting videotex, colour information may be lost.

Table 3.1 Encoded Information Type Conversions (Source X.408)

— Co-operating IPM UA Action. These are service elements which involve the IPM UA in some form of action upon the message, eg receipt notification, auto-forward indication.

— Co-operating IPM UA Information Conveying. These service elements are concerned with the identification of the originator/recipient and status of a message.

The facilities offered by IPM are similar to those provided by existing electronic mail/messaging but with the major advantage of worldwide standardisation and ultimately interconnection. IPM UAs make use of the facilities provided by the MT service and in addition provide other capabilities to form the basic elements of the IPM service. An IPM UA may also perform local functions such as preparation/editing and filing/retrieval of messages. The IPM-UA attaches a header to the message content part/s (see Figure 3.4). This conveys all the information necessary to allow the IPM service to operate. A message may consist of more than one body and each may be of a different content type, eg text, image, voice, etc. When a message has been prepared for transmission, the header and the body part/s will be enveloped by the MT service for despatch to the recipient.

Remote User Agent Access

Usually the MTA and its respective UAs will be collocated, ie implemented on the same system. In this case, the implementation of the interface between these elements is left for the implementor to devise as it will have no effect on the standardised message transfer process. In this way it does not matter about the internal behaviour of the system so long as its external behaviour, the interface between MTAs and from UA to user, is consistent.

Provisions are made, within the model, for the option of remote UA operation, remote meaning that the MTA and UA are not collocated. In such cases the remote or standalone UA is split into two functional entities:

— the User Agent Entity (UAE);

— and the Submission and Delivery Entity (SDE).

The UAE provides the user with the normal UA facilities and functions, while the SDE provides the UAE with access to the facilities of the Message Transfer (MT) service. Whereas message transfer between MTAs is based on the store-and-forward principle, SDE and MTA interaction is interactive in the same way that direct UA to MTA interaction is. This approach allows implementors to have the maximum flexibility when producing their system architecture based on these recommendations.

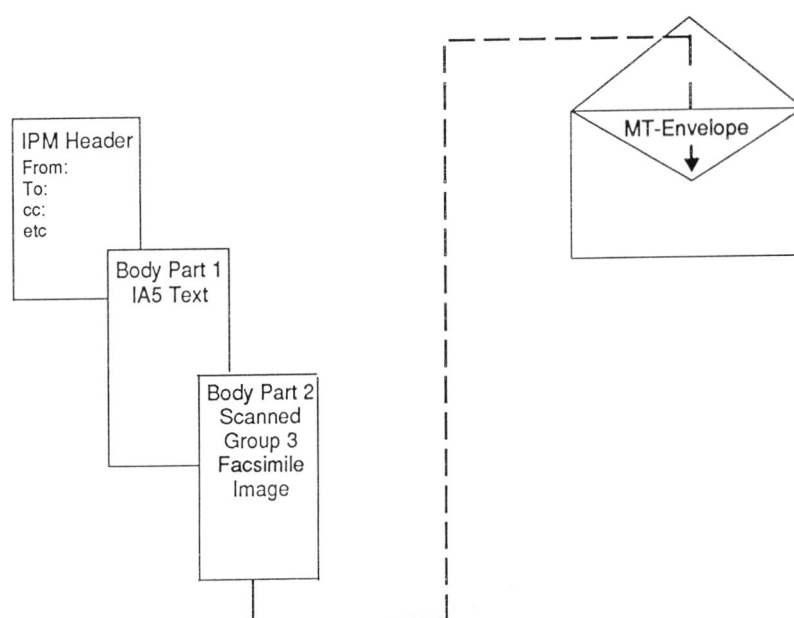

Figure 3.4   The IPM Message Structure

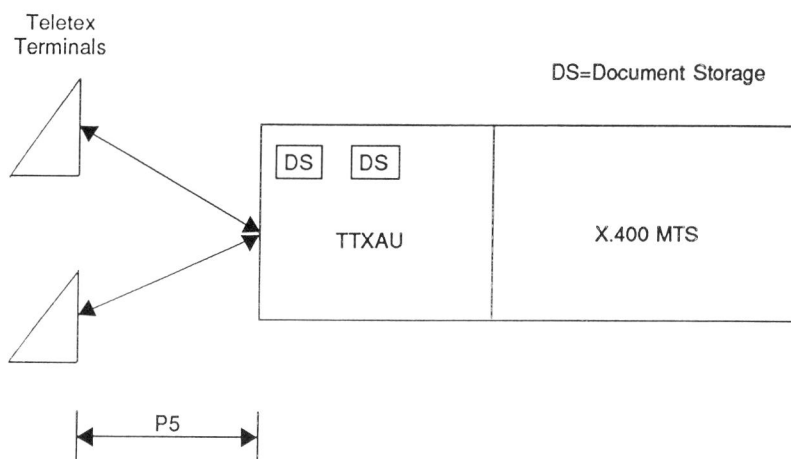

Figure 3.5   TTXAU (Teletex Access Unit)

Teletex Access

Access to systems/services other than teletex from X.400 will be provided via gateway facilities. The functionality of such facilities is not standardised as generally their operation will be – or should be – invisible to the users. Teletex was deemed to be a special case in that it is also a standardised service in which the terminal types, computer-based, have sufficient functionality to make use of the facilities provided by the Interpersonal Messaging Service (IPMS). And at the time that the X.400 recommendations were being devised, it was felt that teletex would have a great future and quickly establish a large user base. In practice this has not been the case, except in Germany — its country of origin.

Teletex access is defined in recommendation X.430 and uses the concept of an access unit, naturally known as a teletex access unit or TTXAU for short (see Figure 3.5). The purpose of the TTXAU is to aid the user of a teletex terminal in gaining access to the features of the IPMS. Basically the TTXAU is an intelligent gateway facility which appears as an IPM-UA to the MT service while providing an interface to teletex on the other side. The TTXAU may also provide Document Storage (DS) which allows messages received from the Message Transfer System (MTS) to be stored under a user's ID until accessed by the teletex user. The teletex terminal and the TTXAU communicate via a protocol known as P5.

The lack of popularity of the teletex service means that the likelihood of TTXAU implementation is fairly low in most countries. However, where it is deemed necessary this option is adequately catered for.

Protocols Used Within Message Handling Systems

Four MHS protocols were defined within the X.400 recommendations, P1, P2, P3 and P5, all of which, except for P5 (teletex access protocol) exist at the application layer of the OSI model (see Appendix 3). The functions of these protocols are as follows:

— P1 provides the basic message relay envelope function between MTAs.

— P2 exists between UAs for the provision of the IPMS. It may be conceived as a standard message header format. P2 is the first of the so-called 'Pc' protocols (or content protocols) each of which is provided along with a new class of UA for particular user applications.

— P3 exists between a remote UAs Submission/Delivery Entity and MTA.

— P5 provides a teletex terminal with access to a teletex access unit (TTXAU) and hence to the facilities of the IPMS.

Reference to Figure 3.6 indicates how these protocols interact with the elements of a message handling system.

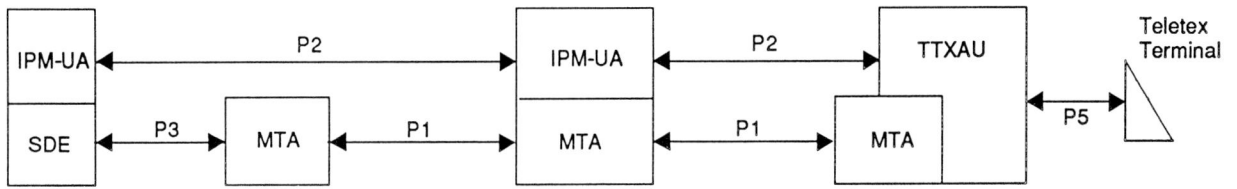

Figure 3.6   Application of P1, P2, P3 AND P5 Protocols

X.410 – Remote Operations and Reliable Transfer Server

Remote operations are used to structure interactive application layer protocols such as the Submission and Delivery Entity (SDE) which uses the P3 protocol. Basically, remote operations specify a structure which allows an operation, in this case Message Transfer (MT) facilities, to be evoked and results from them returned. Remote operations are conveyed over the Reliable Transfer Server (RTS) and hence an OSI network.

The RTS provides the MTA or SDE with a simplified interface to the OSI session layer (layer 5) which is the layer responsible for establishing and maintaining a relationship between user applications wishing to exchange information (see Figure 3.7).

Management Domains Within Message Handling System

After defining the components and services of a message handling system the next stage is to look at how they will be organised and who will have responsibility for them? In a practical message handling environment the systems will be organised into management domains and specific organisations or groups will be responsible for their operation and maintenance.

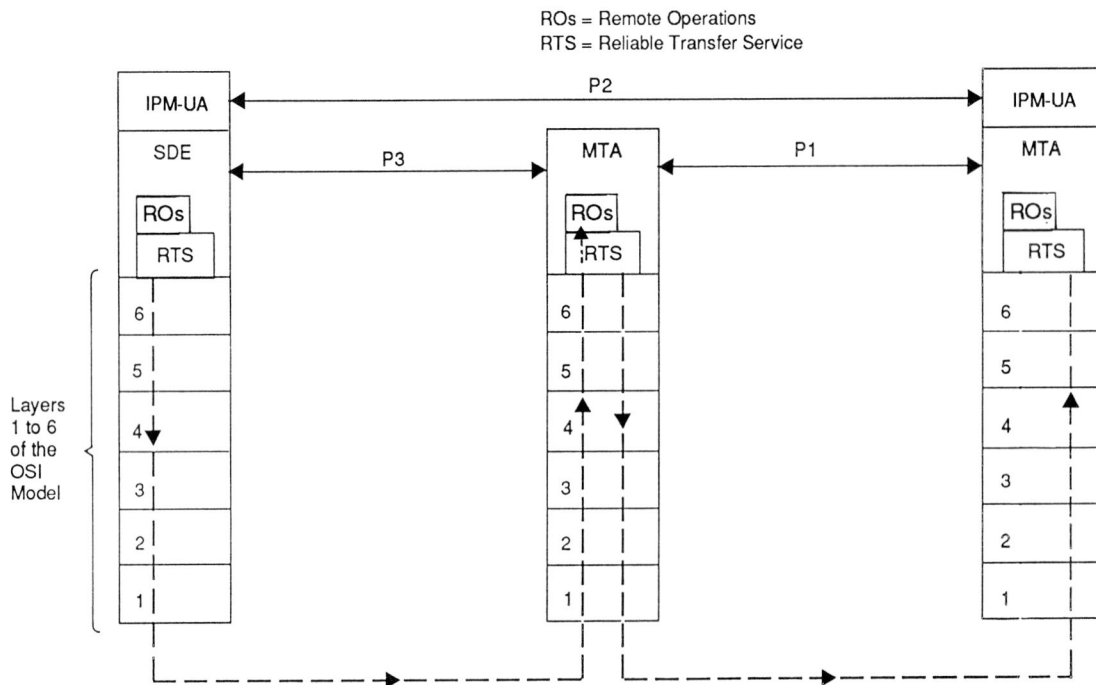

Figure 3.7   The MHS Model Including Remote Operations and the Reliable Transfer Service

Within the X.400 recommendations there exist two types of management domain: Administration Management Domains (ADMDs); and Private Management Domains (PRMDs).

— *Administration Management Domain (ADMD)*. These will form the backbone of the messaging infrastructure by providing links between PRMDs and will be operated by the PTTs. Because of this role they have to provide the majority of optional features defined in the recommendations and in this way they give an added value service. These domains can, also provide connections for individual X.400 users as well as the interconnection between PRMDs. ADMDs may also provide the gateways and access units to other systems and services (see Figure 3.8).

— *Private Management Domains (PRMDs)*. Localised systems which are not under the control of an ADMD will be organised as PRMD, administred by private organisations. Any group can establish a PRMD in order to provide a messaging service to the members of that group. Interconnection between PRMDs is achieved in one of two ways: via connection to an ADMD service; or by direct connection over data circuits. It is expected that the ADMD service will be used to link PRMDs which do not have a large amount of traffic, whereas the direct communication method will be employed where the traffic between PRMDs is sufficient to justify it. The PRMDs will provide a more specialised, but not necessarily limited service which is tailored to the individual requirements of an organisation.

Figure 3.9 gives some indication of how these domains might be interconnected both internally within one country and on an international basis between two countries. This is the CCITT view of connections and therefore they do not see the potential or need for direct inter-PRMD links. ISO's MOTIS work does allow such connections and since this work forms the basis of the European CEPT and CEN/CENELEC profiles (see Appendix 3, 5, and 7) it is a worthwhile subject to mention. In addition the MOTIS work also specifies other features for private messaging systems such as intra-domain trace information useful in the administration of a PRMD.

X.400 Naming/Addressing

In the past, people have struggled to come to terms with the non-user-friendly addressing schemes employed within electronic messaging systems. In a deliberate move against this potential obstacle to user acceptance it was decided to attempt to recify this situation in X.400 systems. Hence the concept of the Originator/Recipient (O/R) name and address was introduced. Before discussing these further it is important to specify the difference between an X.400 name and address:

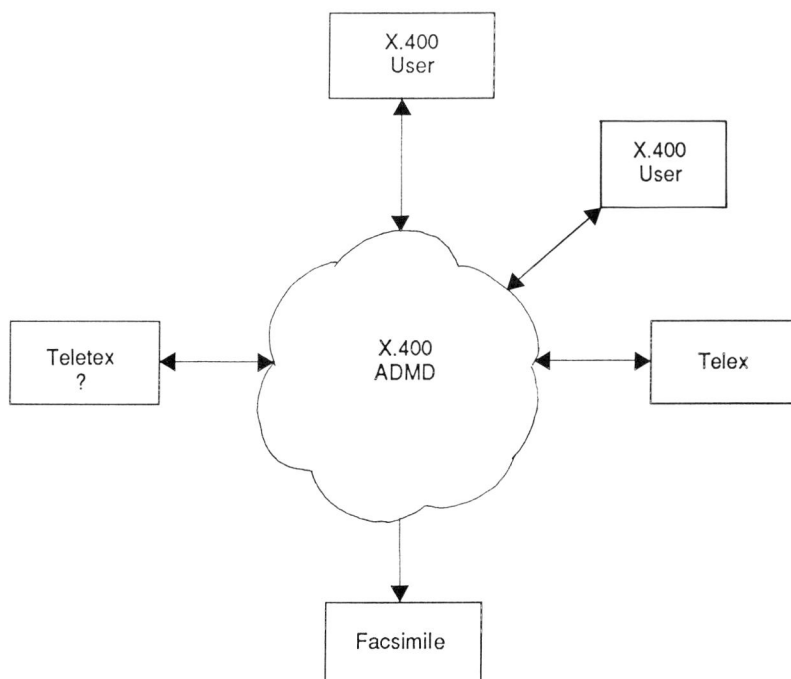

Figure 3.8   Typical Facilities of an ADMD

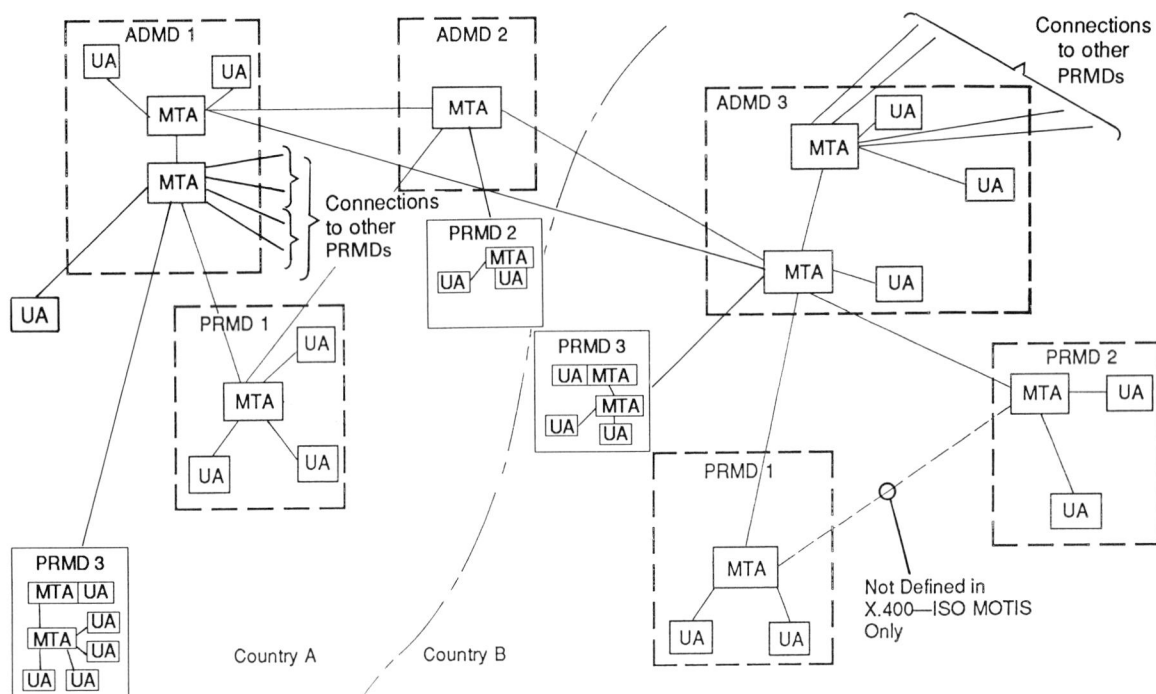

Figure 3.9   Interconnection of Administration and Private Domains

— An O/R name is a descriptive name of the actual human user which is applied to their personal UA and thus defines an entity by specifying what it is. Such a name can be used to specify the intended recipient of a message but would require the MTS to perform a directory look-up function in order to determine where the recipient is located, eg which ADMD, PRMD, etc.

— An O/R address is a descriptive name for a UA which has certain characteristics to help the MTS locate the UA within the messaging environment. Every O/R address is an O/R name, but not every O/R name is an O/R address, because the O/R name may be incomplete. The use of an O/R address to identify the intended recipient of a message implies that the MTS will not have to use the directory as it has already been given the required information. This feature will be very useful in the early stages of X.400 MHS implementation before the arrival of a full international directory service.

It is the intention that when using the concept of O/R names a message originator will be able to provide a descriptive name for each recipient using information commonly known about that user. Such information might be that which is typically found on a business card, the idea being that just enough information has to be given in order to uniquely specify the recipient. There are four categories of standard attributes for an O/R name which have been identified so far. These categories, along with the attributes supported by MHS protocols, are defined below:

Personal Attributes

     personal name.

Geographical Attributes

     country name.

Organisational Attributes

     organisation name;
     organisational unit.

Architectural Attributes

> X.121 address;
> unique UA identifier (only numeric values);
> administration mangement domain (ADMD) name;
> private management domain (PRMD) name.

(Note: X.121 is a CCITT recommendation defining an international numbering plan for public data networks.)

This is the list of attributes specified in the 1984 version of X.400, but it may be expanded to encompass others after future development. The MHS protocols also support Domain Defined Attributes (DDAs) which allow for the transmission of extra addressing information, not standardised within X.400, with the message, which can be used privately within a domain eg to allow a message to be routed through domains which are not X.400 compatible.

At present there are two forms of O/R address which have been defined within the X.400 recommendations. Form 1, in one of its three variant forms, is meant to identify MHS users only, and Form 2 is intended primarily to identify users of teletex and other telematic services, such as telex and facsimile. These two forms of O/R name are detailed below.

FORM 1

Variant 1
O/R address consist of: Country name
                            Administration domain name
                            [Private domain name]
                            [Personal name]
                            [Organisation name]
                            [Organisational unit names]
                            Domain defined attributes

(Note: At least one of the attributes within [], with the exception of DDAs which are optional, must be selected in addition to the first two.)

Variant 2
O/R address consists of: Country name
                            Administration domain name
                            UA unique numeric identifier
                            [Domain defined attributes]

(Note: DDAs optional.)

Variant 3
O/R address consists of: Country name
                            Administration domain name
                            X.121 address
                            [Domain defined attributes]

(Note: DDAs optional.)

FORM 2

O/R address consists of: X.121 address

                            [Telematic terminal identifier]

(Note: Telematic terminal identifier is optional.)

The eventual aim within X.400 systems is that users will be able to address messages with the O/R name. The MHS will then search through directories (implemented in accordance with the CCITT X.500 directory recommendations – see Appendix 4) until a unique O/R address is located. This O/R address will be used to correctly route the message across the MTS. However until directories compatible with X.400 are developed the simple O/R address will need to be used to identify the intended recipient for a message. Only enough information need be included in the O/R address for it to be correctly delivered, ie sufficient to uniquely identify the recipient (eg a user named Smith will undoubtedly require more attributes for identification than someone with a less common surname). Once X.500 directories are developed it is hoped that they will be interconnected to provide a worldwide directory service.

4   The UK Messaging Infrastructure

REGULATORY FRAMEWORK IN THE UK

In recent years there have been drastic changes in the regulatory framework surrounding British Telecom (BT) and its monopoly position in the provision of telecommunication and data services in the UK. Prior to the liberalisation of BT in 1981 there existed a monopoly situation. Post Office Telecommunications (now BT) derived much of its power from its exclusive carrying privileges. This gave the organisation sole rights to transmit information by a wide range of physical means. Any use of BT circuits which may have been outside the conditions laid down were required to be licensed. Few exceptions were authorised, and many data communications applications had to have licensing approval.

Licences were issued for private circuits and networks on the condition that they were solely for the use of the licensed party within his own organisation. Third parties could not use the networks or services carried by them. Tight control was also kept on connections between the public switched telephone network (PSTN) and private circuits.

This heavily controlled environment had existed for some considerable time when, in the 1970s, it became apparent that this was stifling innovation and the development of new IT applications. It was at this stage that the government initiated a liberalisation programme aimed at relaxing the monpoly by removing some of the restrictions and introducing a greater element of competition into the supply of tele-communications products and services. These developments were eventually to transform BT – previously a government department with responsibility for the provision of UK telecommunication services – into a private company operating under a government licence for the provision of services in competition with others.

The first stage of the liberalisation of the UK telecommunications market began with the 1981 British Telecommunications Act. From this date the government could license equipment and service providers to operate in direct competition with BT. It was at this time that BT set up the British Approvals Board for Telecommunications (BABT 'green dot' approval logo) to test the compliance of Telecom's equipment to approval standards. BABT approval is required for any vendor's equipment which is to be connected to BT's network.

The 1984 Telecommunications Act marked the beginning of BT as a public limited company and 51 per cent of its then government-owned shares where subsequently sold to the public. The Act also marked the end of BT's exclusive right to run public telecommunications systems and it began operating under a government-granted Public Telecommunications Operator's (PTO) licence along with Mercury (licensed in 1982) and Kingston-upon-Hull City Council (which has been running a telephone service since 1904).

In order to ensure the smooth running of the UK's telecommunications services, the government also established a new organisation, the Office of Telecommunications (Oftel). Oftel, headed by the Director General of Telecommunications (currently Professor Bryan Carsberg) has responsibility for overseeing the telecommunications market in the UK, safeguarding the interests of the consumer and making sure that the conditions of the licences are adhered to. The Director General of Oftel can enforce the licence conditions through the courts if necessary. He also has the power to modify the conditions of the licence, either with the agreement of the licensee, or after reference to the Monopolies and Mergers Commision.

The above effectively defines the regulatory position in the UK market with respect to the provision of telecommunication and data services; but in order to ascertain how this framework will affect the growth of the UK X.400 messaging infrastructure a specific type of licence has to be examined.

'VALUE ADDED' SERVICES LICENCE

In 1983 the government established the VANS General Licence which enabled the provision of Value Added Network Services (VANS). Under the conditions of the VANS General Licence companies were allowed to re-sell BT circuits as long as they were providing 'added value'. This 'added value' is deemed to occur when the information or message being transferred is modified or acted on in some way. This action could take any of the following forms:

— the message could be stored by the Value Added operator for retrieval at a later time or for forwarding to specified destination(s);

— protocol conversion could be performed;

— speed conversion could also be performed;

— the content of the message could be changed to reflect some current information.

The nature of the intended service had to be specified by the licence applicants, and a fee was levied. Many services have developed based on this licence, so that today Value Added operators are quite common.

In the period since its issue it was clear that the VANS General Licence failed to meet the government's criterion in certain respects. This resulted in a review of the existing document and the eventual publication of a new one in 1987. The new licence appeared under a different name, the Value Added and Data Services or VADS licence. The VADS licence is available to anyone except:

— those who have had the licence revoked;

— the PTOs and their associates. The PTOs are licensed to provide VADS under their PTO licences.

It is not necessary to apply for a licence, as it is a class licence (ie assumed to be granted to all, unless revoked). However, if either of the following conditions applies the licensee is required to register the company's name and address with the Director General of Telecommunications (the Director of Oftel):

— if the licensee is a major service provider (where the company's or group's turnover from VADS exceeds £1 million a year or where the group's turnover from all activities exceeds £50 million a year);

— if the licensee is a trilateral service provider, that is, where switching services are provided between the public network and another company outside the licensee's group, over a line leased from a PTO, or between two such companies over leased lines (see Figure 4.1).

From the description given it should by now be obvious that the VADS licence would have jurisdiction over X.400 Administration Management Domains (ADMDs) or X.400 services which will develop in the UK. This would be true in all cases except in the that of PTOs, such as BT, who are licensed under their PTO licence to provide such services.

There are general conditions to which the holder of a VADS licence must adhere to, and these are detailed within the licence. One of these conditions in particular has a crucial bearing on message handling and indeed the future coherent development of a UK messaging infrastructure: it is the numbering condition. A licensee is required to adopt a plan for the allocation of numbers or addresses to subscribers to their service. This plan must be approved by the Director General of Telecommunications to ensure that there is no overlap with numbers or addresses used by other networks, which could cause problems such as misrouteing of messages. This is one area in which the development of a messaging infrastructure in any country requires strict control.

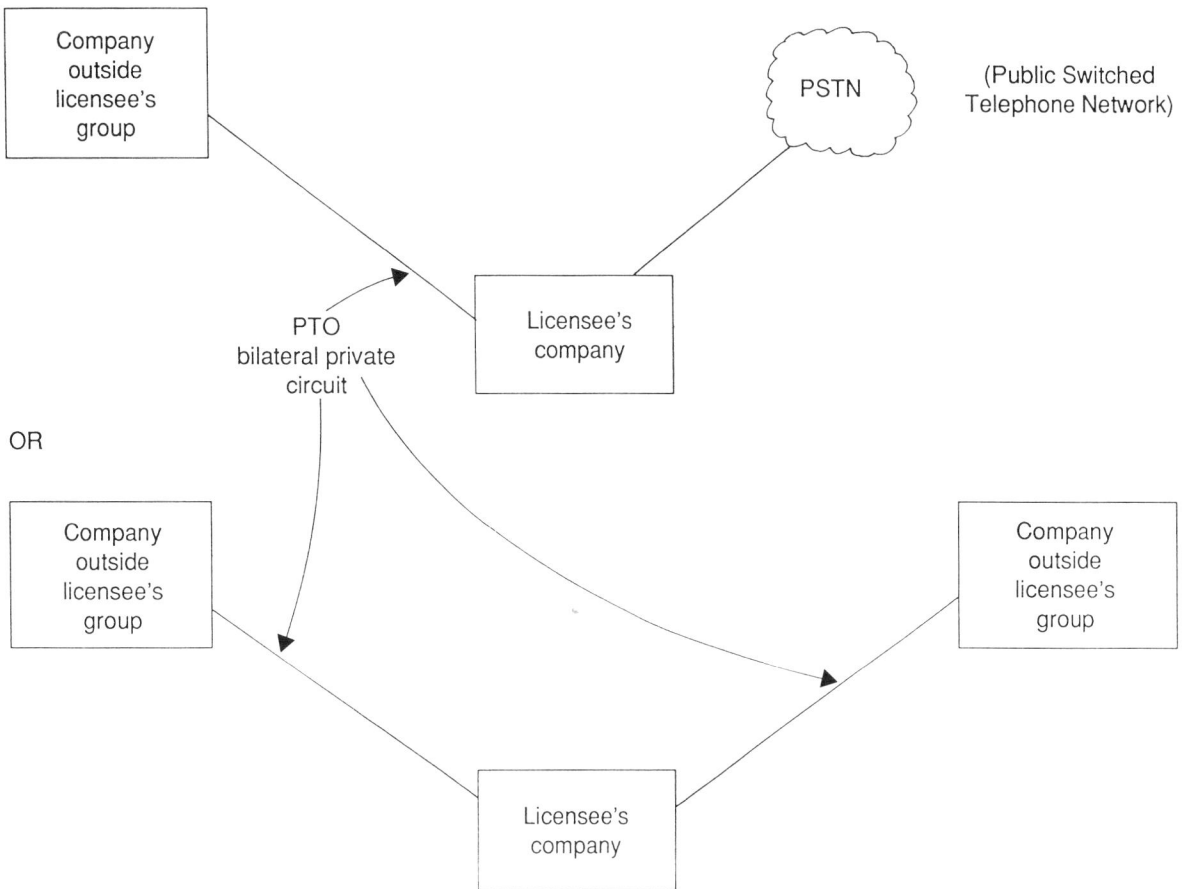

Figure 4.1   Trilateral Service

BRITISH TELECOM'S GOLD 400 MESSAGE HANDLING SERVICE

Although it is some time since British Telecom lost its monopolistic position as a provider of communication services, they are still by far the largest operator in the UK. With this position they were expected to be a major provider of message handling services. This expectation was confirmed in late 1985 with BT announcing their intention to provide a message handling service and products conforming to the CCITT's X.400 recommendations. This announcement, as recent events have proved, was some way in advance of the provision of a real service and products. BT's reasoning was that by stating their intentions as early as possible they would instill sufficient confidence in the computer industry to allow the development of conformant products to yield connection to the service when it arrived.

One of the first steps was to specify the functionality to be provided by the service in order to give a clear picture of what to expect when connecting to the future message handling service. This involved producing a profile (see definition Appendix 5) of the X.400 recommendations which would be used as a basis for the service. This document is one from the series , entitled Open Network Architecture (ONA), which BT are producing to cover all aspects of their OSI implementations. ONA is BT's strategy for implementing international standards across its whole range of IT products and services, and Gold 400 is the first ONA service. As such, it will form the core of a range of ONA conformant products and services, all of which are guaranteed to interwork.

The BT MHS service is based on software from the BT US subsidiary, Dialcom. The Dialcom 400 software, used for Gold 400, runs on Prime hardware. Recently BT announced a major restructuring of its data services activities in an enlarged Dialcom Group whose range of products and services will include:

— Dialcom's electronic mail, messaging and information services, plus their X.400 software— Dialcom 400;

— Telecom Gold and Gold 400;

— Value-Added Business Services, including Prestel.

In the autumn of 1986 BT commenced engineering trials in collaboration with several X.400 vendors whose implementations were sufficiently advanced and could be tested against the CEPT A/311 profiles.

The BT vendor's forum, on 25 June 1987, marked the first public statements about how their message handling service would appear in its final form. The name of the service, Gold 400, was announced, along with dates for the commencement of the pilot trial, the full service, the progressive introduction of other facilities (such as telex access) and the schedule of the charges (detailed later). At this event some notable vendors (DEC, Data General, Hewlett Packard, ICL and BT's own products division) also detailed their X.400 conformant products.

BRITISH TELECOM'S INVOLVEMENT IN INTERWORKING DEMONSTRATIONS

British Telecom, along with Deutsche Bundespost, were the only service providers to be involved in the ceBIT '87 X.400 demonstration (see Figure 4.2). This was one of the first really meaningful demonstrations because all the participants had to undertake that they would provide X.400 conformant products or services before the end of 1987. Because of this it was not just a demonstration of prototypes, but rather of actual products and services showing a realistic idea of PRMD/ADMD interworking.

At the Networks '87 exhibition at Wembley in June 1987 BT displayed their X.400 service for the first time under its Gold 400 name. At this event, BT and DEC collaborated to provide reciprocal demonstrations on each other's stands.

Telecom '87 was the major event in the X.400 calendar for 1987 and again BT were involved along with 20 other service providers and product vendors. This event was most notable for its interconnection of nine

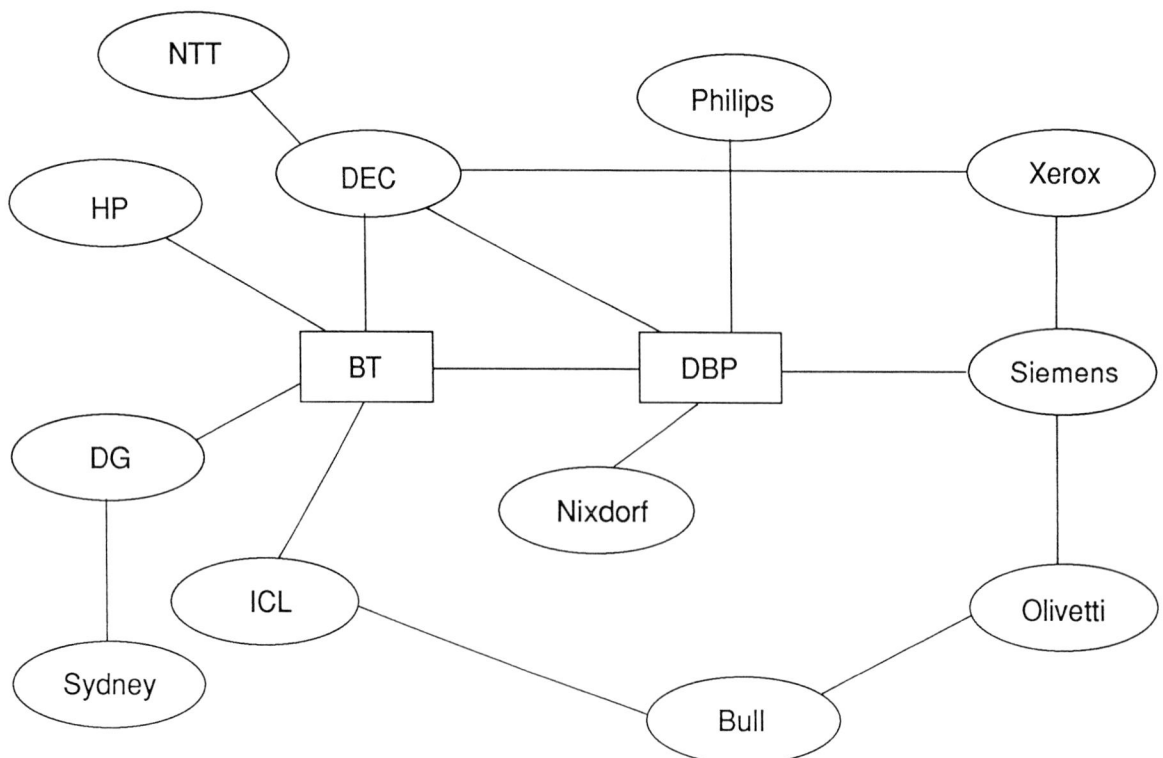

Figure 4.2   The ceBIT 1987 X.400 Demonstration

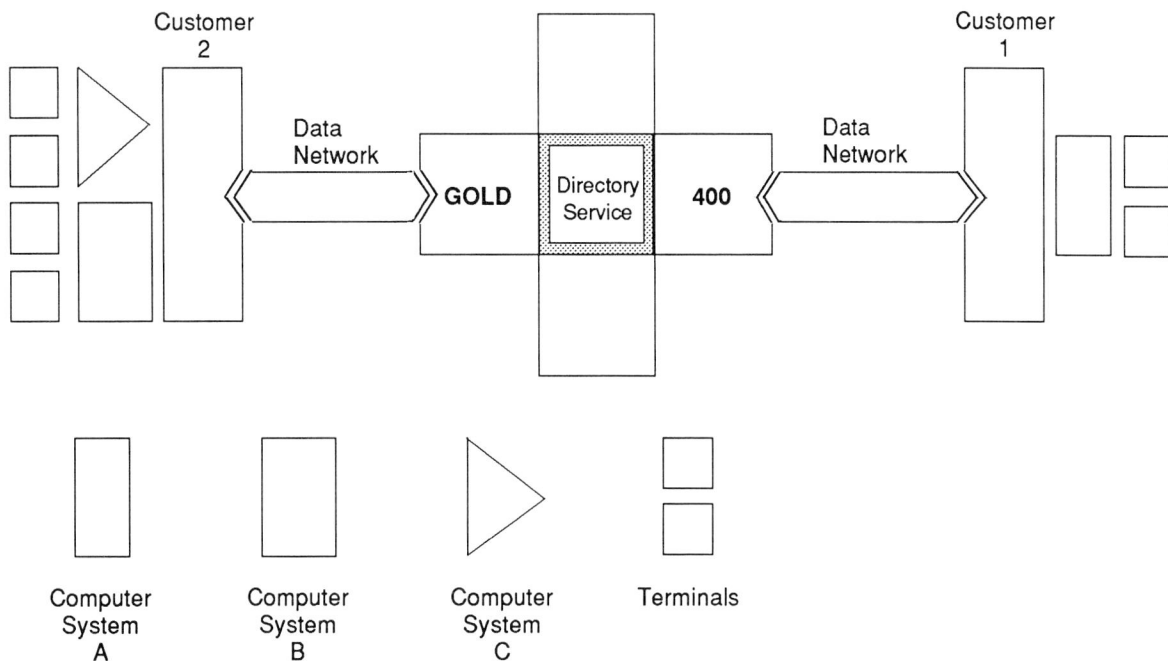

Figure 4.3   The Gold 400 Pilot Trial Phase

X.400 services, this platform giving a strong indication of some of the links to be established in the future which will allow an X.400 user access to all corners of the globe.

BT are also members of the EurOSInet group (see Chapter 7) who demonstrated X.400 interconnection at the Compec '87 exhibition. In line with the general aims of the EurOSInet group, this demonstrator is to be made available for use at other events in the future.

For further information about all of these interworking demonstrations, with the exception of Networks '87, refer to Chapter 7.

STAGES IN THE PROVISION OF GOLD 400

(Author's Note: Please note that the dates given in this section for the provision of Gold 400 facilities are the best available at the time of writing and may be subject to change at a later date.)

On 1 July 1987 Gold 400 was launched in its pilot trial phase, which marked the end of the engineering trials carried out since autumn 1986. During this phase BT allowed organisations to become subscribers to the service, on a pilot basis, through payment of a registration fee but not being charged for the traffic carried by the service. During this period BT were actively seeking feedback from the users participating in the trial.

The pilot trial was on the basis of PRMD connection (P1) to Gold 400. In this situation the service acted as a relay agent between PRMDs (see Figure 4.3). Connection to the service was via BT's Packet Switch Stream (PSS) service and a such products connecting to this must have PSS PTC approval (as required by Section 22 of the 1984 Telecommunications Act). The only other option offered during this period was the on-line directory service, access to which is gained via the Telecom Gold electronic mail service. The directory service is an 'interactive' facility; it is just a special type of database service, and hence cannot be accessed and used over the Gold 400 connection. The connection charge to Telecom Gold is covered by the appropriate Gold 400 registration fee, while access to the directory service is charged for at normal Telecom Gold rates. During this phase these were the only facilities provided by Gold 400.

July 1988 was the date quoted for the provision of the Gold 400 commercial service, see Figure 4.4. At this stage the list of facilities provided should be as follows:

— *Relay.* Identical to that of the pilot trial, relay of messages between PRMDs.

— *Link to Telecom Gold.* This will mean that Gold 400 users and Telecom Gold users (135,000 public subscribers) will be able to exchange messages. The connection between the services centres

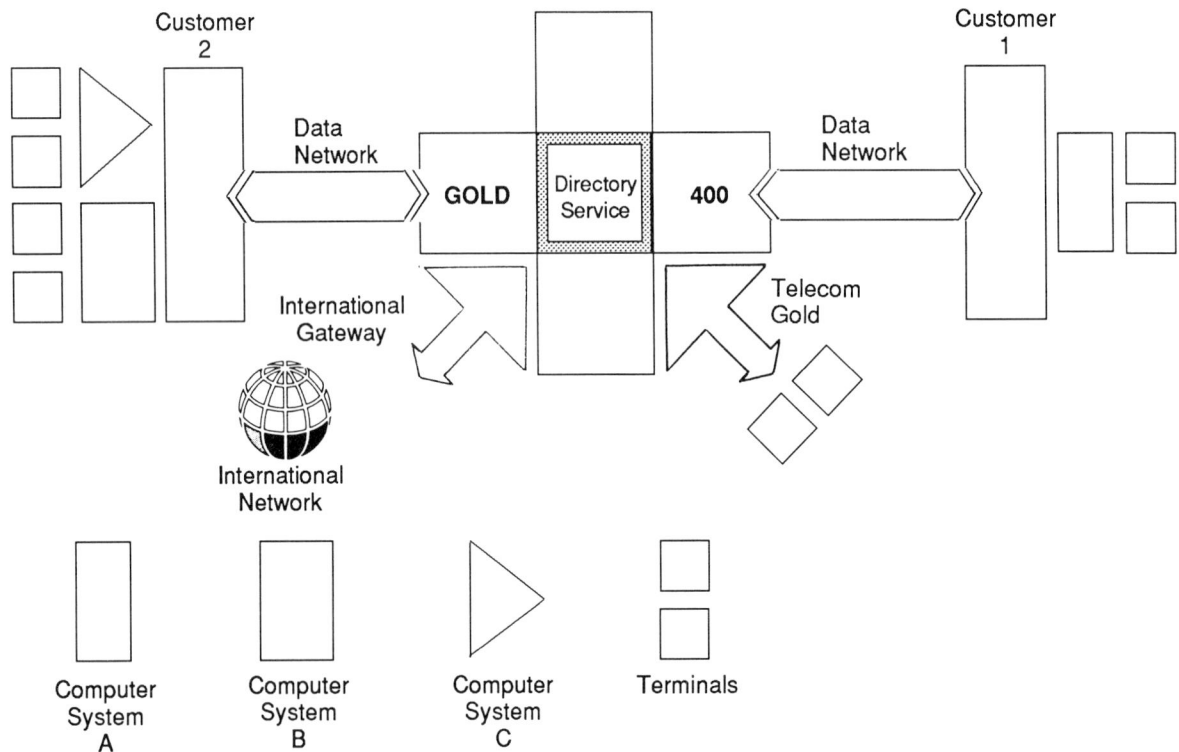

Figure 4.4   April 1988 Full Gold 400 Service

around the use of the interactive directory, available to both services, which will yield sufficient information for routeing messages in either direction. It is intended that in the future (mid 1989) the software for the standard Telecom Gold e-mail service will be uprated to full X.400 User Agent (UA) compatibility; but for the interim a gateway between the services is sufficient.

— *Directory service.* Available to both the users of Gold 400 and Telecom Gold.

— *International links.* International links to Gold 400 are now in place to other services in the USA and the Netherlands. It should also be remembered that the interconnection platform, for the PTTs and service providers, utilised in the Telecom '87 exhibition, still exists.

The option of access from Telecom Gold is important as it allows people and organisations with only dumb terminals, portable PCs, viewdata and PC terminals to gain simple dial-up access to X.400 messaging facilities.

From July 1989 the Gold 400 service (see Figure 4.5) should be able to exchange messages with the worldwide telex network (approximately 1.5 million terminals) and deliver messages to facsimile terminals (approaching the same numbers as telex). These facilities have been publicly demonstrated and are now subject to trials by current Gold 400 users.

Delivery from Gold 400 to telex will be achieved by means of a conversion facility which will carry out the necessary operations on the message in order to make it compatible with telex terminals. Delivery of a telex message to a Gold 400 user is a little more complex and different options are being evaluated at present, one of these is:

— *Direct Dialling In (DDI).* In this case each Gold 400 user would have a specific number to yield access from a telex terminal. This information would probably be carried on a business card along with the more normal X.400 address.

Also under consideration for provision at the end of 1989 is a physical delivery service from Gold 400.

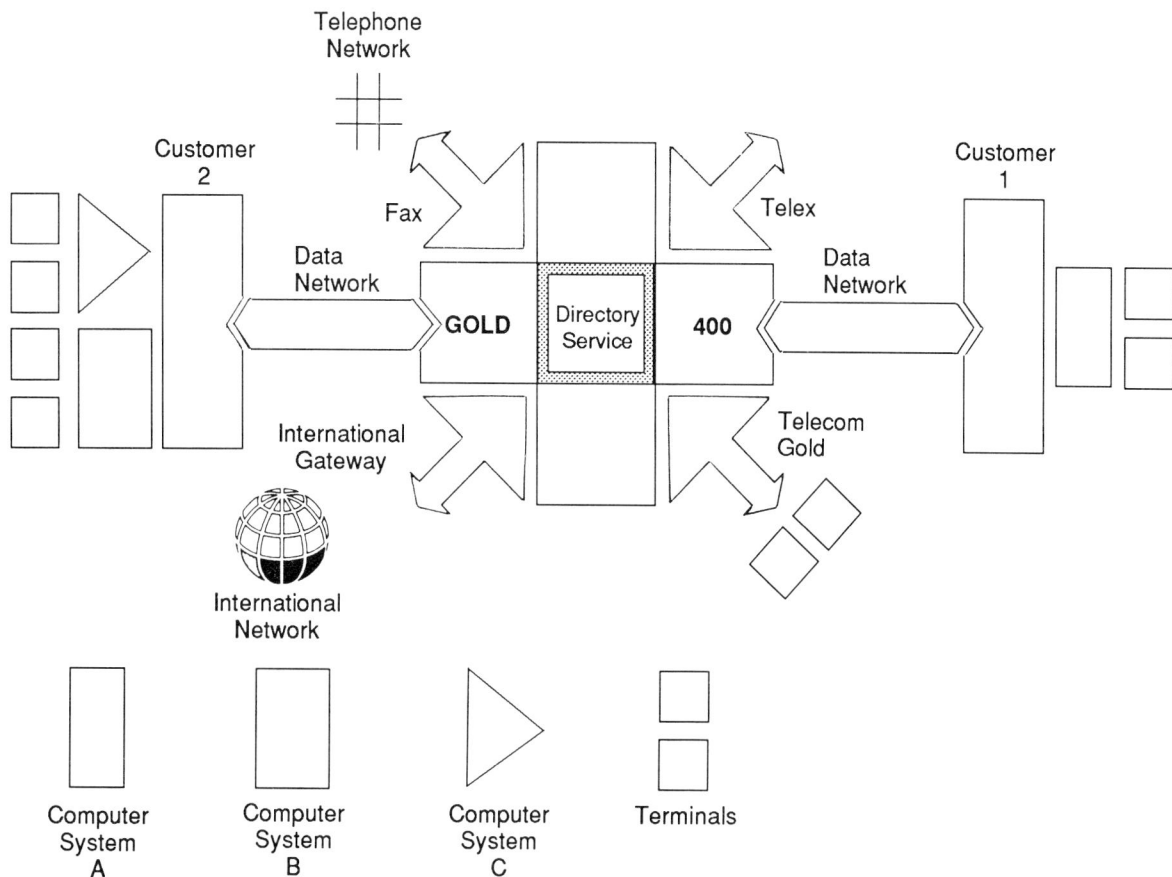

Figure 4.5   Gold 400 with the Addition of Telex and Facsimile Access

This would allow a Gold 400 user to have a message delivered by services such as the Royal Mail and couriers, to account for the situation where the recipient is not connected to the service or any other electronic messaging media. The use of this service would naturally require that the message had to include a full and correct postal address.

Some other interesting Gold 400 features also being considered for provision in 1989/1990 are:

— P7 remote User Agent (UA) and Message Store support, see Appendix 7 for explanation of these concepts;

— Revisable Document Exchange (RDE) and Document Architecture Conversion (DAC)/ODA (Office Document Architecture);

— Electronic Data Interchange over X.400;

— 1988 X.400 extensions (see Appendix 7 for full explanation of these).

The main application area for Gold 400 at present is intercompany IPMS (interpersonal messaging) by the interconnection of different public and private messaging services and systems. In the future BT hope that Gold 400 will act as a Reliable Transfer Service (RTS) platform to support the growth of traffic from industry specific applications requiring the transfer of data between incompatible hardware, eg CAD/CAM, data files, EDI, EFT, etc.

Getting Started With Gold 400

The following three documents are British Telecom's interim attachment policy, registration form, and charges schedule, respectively for their Gold 400 message handling service (see Figures 4.6, 4.7, 4.8). These are the first items to look at when considering registering for the service.

BT Policy For Attachment To GOLD 400

Purpose: This document defines how Users of X.400 products requiring attachment to GOLD 400 will achieve compatability with the public service requirements.

1   The product must already have approval under section 22 of the Telecommunications Act 1984 for connection (directly or indirectly) to the BT system.

2   The User of an X.400 product who requires attachment to GOLD 400 must register with Dialcom's GOLD 400 Service Administration. The GOLD 400 registration process will include payment of the published registration fee.

3   The User of an X.400 product who requires attachment to GOLD 400 will be required to provide, on registration, a declaration from the X.400 product Vendor that the User's implementation of X.400 is compliant with BT's requirements, ie with the CEPT ENV 41202 Profile of X.400.

4   It is strongly recommended that Vendors submit their X.400 product for test with an appropriate X.400 conformance test house prior to any attachment of an implementation of their X.400 product to GOLD 400. It will not be mandatory, however, for the User of a Vendor's X.400 product to submit their X.400 implementation for a pre-Registration conformance test.

5   In the case of problems with a Customer's X.400 implementation which affect GOLD 400, BT will initially advise the Customer that problems with the implementation are affecting Service, and may issue an instruction in accordance with paragraphs 4.1.5 and 4.2 of the Conditions for Telecom Gold Service under which GOLD 400 is provided, that these problems must be rectified within a specified timescale and the rectification certified to BT. Failure to comply with such an instruction may result in BT taking the action described in 6 below. (An example of such a case would be where an implementation jeopardised GOLD 400's ability to comply with the responsibilities assigned to an Administration Management Domain in X.400 84, 2.3.2.4.)

6   As provided for in paragraph 11.1.3 of the Conditions for Telecom Gold Service, BT reserves the right, in the case of failure on the Customer's part to comply with the Conditions (which would include compliance with an instruction issued under paragraph 4.1.5 of the Conditions) to suspend Service to the Customer or terminate the contract. As indicated in paragraph 5 above, any instruction would require rectification to be certified to BT. BT would require a certificate from an X.400 conformance test house that the Customer's X.400 implementation conforms to the CEPT ENV 41202 Profile of X.400.

Notification to the Customer of BT taking action under these circumstances would clearly state that this was due to the Customer failing to comply with an instruction to rectify a failure to meet the required standard for continued attachment to GOLD 400. (For information: BT Teleprove is an independent X.400 consultancy service and product conformance test house).

7   As required by Condition 23 of the British Telecommunications Licence, when certain changes are made to BT's networks and services, advance notice will be given to those likely to be affected, in accordance with procedures published in consultation with the Director General of Telecommunications (Oftel). Copies of these procedures may be obtained from a BT District Office.

GOLD 400
Dialcom
Direct Response Unit
PO Box 1351
London NW2 7HZ
Telephone: 0800 200 700

A member of the Dialcom Group, worldwide supplier of electronic messaging and value-added data services,
British Telecommunications plc, Registered Office: 81 Newgate Street, London EC1A 7AJ. Registered in England
No. 1800000.

Telecom Gold and Gold 400 are trademarks of British Telecommunications plc.   TG/G4/AA1/5.88

Figure 4.6   BT Attachment Policy Statement

A P P L I C A T I O N   T O   R E G I S T E R
W I T H   B R I T I S H   T E L E C O M M U N I C A T I O N S   P L C   (B T)
F O R   T E L E C O M   G O L D   S E R V I C E

GOLD
400

Please complete in Block Capital Letters

Customer/Company Name

1 ☐☐☐☐☐☐☐☐☐☐☐☐☐☐☐☐☐☐☐☐☐☐☐☐☐☐☐☐☐☐☐☐☐☐☐☐☐☐

Nature of Business

2 ☐☐☐☐☐☐☐☐☐☐☐☐☐☐☐☐☐☐☐☐☐☐☐☐☐☐☐☐☐☐☐☐☐☐☐☐☐☐

Customer Address

3 ☐☐☐☐☐☐☐☐☐☐☐☐☐☐☐☐☐☐☐☐☐☐☐☐☐☐☐☐☐☐☐☐☐☐☐☐☐☐

Postcode   County
☐☐☐☐☐☐☐☐☐   ☐☐☐☐☐☐☐☐☐☐☐☐☐☐☐

Telephone

4 ☐☐☐☐☐☐☐☐☐☐☐☐☐

Telex   Fax
☐☐☐☐☐☐☐☐☐☐☐☐   ☐☐☐☐☐☐☐☐☐☐

Contact Name

5 ☐☐☐☐☐☐☐☐☐☐☐☐☐☐☐

Position
☐☐☐☐☐☐☐☐☐☐☐☐☐☐☐

6 If you are already registered for Telecom Gold Electronic Mail Service then please state:

Corporate/Club Account   System ☐☐   Mailbox Prefix ☐☐☐

Dialcom Sales Contact

7 ☐☐☐☐☐☐☐☐☐☐☐☐☐☐☐☐☐☐☐☐☐☐☐☐☐☐☐☐☐☐☐☐☐☐☐☐☐☐

British Telecom National Account Manager (where applicable)

8 ☐☐☐☐☐☐☐☐☐☐☐☐☐☐☐☐☐☐☐☐☐☐☐☐☐☐☐☐☐☐☐☐☐☐☐☐☐☐

I apply for GOLD 400 on BT's Conditions for Telecom Gold Service, and according to the Attachment Policy overleaf, which I accept. I enclose payment/Please invoice for the appropriate amount, as shown in the current price list, made payable to British Telecommunications plc, bvy way of the registration charge.

Name
☐☐☐☐☐☐☐☐☐☐☐☐☐☐☐☐☐☐☐☐☐☐☐☐☐☐☐☐☐☐☐☐☐☐☐☐☐☐

Position
☐☐☐☐☐☐☐☐☐☐☐☐☐☐☐☐☐☐☐☐☐☐☐☐☐☐☐☐☐☐☐☐☐☐☐☐☐☐

Signed   Dated

When the Customer is a limited company or other corporate body, the form must be signed by a duly authorised person. Partners should sign for self and other partners.

Please complete the rest of the registration information on the following pages.

Figure 4.7   BT Registration Form

C H A R G E S   S C H E D U L E

GOLD
400

Corporate Registration

Corporate Registration of each Private Management Domain (PRMD) for connection to GOLD 400 — 1,000 pounds. This applies to single PRMD name registrations only.

One Organisation Structure entered in the Directory and entries within it under customer control.

Telecom Gold mailboxes for access to the Directory; plus Database gateway services provided at Telecom Gold standard charges.

Initial Directory Structure setup — up to five user entries.

For companies already a Telecom Gold Corporate subscriber, the initial registration charge is reduced to 700 pounds.

Messaging Usage Charges

All usage charges are based on block units of 2048 characters or part thereof.

Messages from PRMD's to Telecom Gold mailboxes (GOLD) are charged at a reduced rate to those sent to other PRMD's (RELAY).

	Relay	Gold
Standard Rate 8am — 7pm Mon-Fri (ex. Bank holidays)	20 pence	16 pence
Off Peak Rate All other times	15 pence	12 pence
Delivery Notification	25 pence	25 pence
Return of contents	15 pence	15 pence
Non-delivery notification	No charge	No charge

Directory Access

Access to Telecom Gold as per standard price list.
Directory access surcharge — per minute or part thereof.   5 pence

Minimum Monthly Billing

Minimum invoice amount for Telecom Gold and GOLD 400 Service per month   100 pounds

Multiple Registrations

Additional PRMD's registered will be charged as follows:
6 — 10 PRMD's   750 Pounds
11 +   PRMD's   500 Pounds

This charge will be levied regardless of the PRMD's being located at the same network address or not.

Each additional Organisation Structure created in the Directory will be charged
at a rate of   250 pounds

All Charges quoted are exclusive of VAT and although correct at the time of print, information shown is subect to revision. June 1988

Figure 4.8   BT Charges Schedule

Teleprove is BT's test house which has been testing telecommuncations products for over 30 years. They are offering a testing service set up in collaboration with British Telecom Research Labs (BTRL) to serve the growing UK X.400 market. The MHS Interworking Test Service is based upon the Danet (a German company) MHTS/400 test system which will allow vendors and customers to check the major X.400 features of their message handling systems.

Explanation of the Charging Scheme for Gold 400

There is only one type of registration on offer:

— connection of a PRMD to the Gold 400 ADMD and registration as a Telecom Gold user which allows access to the directory, electronic mail and database services. All users of Gold 400 can be entered to the directory (number unlimited) and use of Telecom Gold for any of the above services will be charged at the standard rates of this service. Registration fee is £1000, this figure may rise shortly – hence check with BT Dialcom.

The rest of the charges listed are fairly self-explanatory.

The issue which is not dealt with on the charge schedule is the cost of the connection to Gold 400 via the PSS service. These charges are additional to the Gold 400 usage charges. Datalines permanently connected to the PSS allow Gold 400 to be accessed. Either of the following standard dataline connections is acceptable:

— Dataline 2400, giving a line speed of 2400 bps full duplex;

— Dataline 9600, giving a line speed of 9600 bps full duplex.

The charges for these circuits are as follows:

Dataline	Connection charge	Quarterly rental
2400	£850	£525
9600	£950	£925

In addition to these will be the charges made for the calls via the PSS to Gold 400:

Datacalls – Inland Call Charges					
Volume charge per kilosegment			Duration charge per hour		
Standard	Cheap	Low	Standard	Cheap	Low
£0.30	£0.20	£0.185	£0.30	£0.20	£0.185

All the charges detailed so far apply when considering a Gold 400 connection, hence, care should be taken if a clear idea of the total cost is to be gained. The most useful approach is to consider the type and volume of messages to be transferred on a daily basis, work out the average cost and then contrast this figure with the benefits of an X.400 connection. It should also be emphasised that a PSS connection can be used for other PSS traffic in addition to Gold 400, hence it is up to the user to make the best usage of this facility. Certainly Gold 400 is not a cheap service, but in the long term its potential advantages could probably outweigh the cost.

FUTURE DEVELOPMENT OF THE UK MESSAGING INFRASTRUCTURE

For the moment BT's Gold 400 service is the only message handling ADMD facility within the UK and as such it is likely to act as the major backbone of this country's future messaging infrastructure. During 1989 the service will offer connection to telex, facsimile, Telecom Gold and other international X.400 message handling services. These connections are all added value features which should help to show users that an X.400 connection is a very worthwhile business tool. Gold 400 currently has 60 attached private networks with potential contact to thousands of corporate users.

Within the UK there is also a growing band of users who are connected to public electronic mail services, such as: Mercury Link 7500 (was Easylink), Quik-Comm (GE IS), Comet (ISTEL) and One-to-One. The individual user bases of these services cannot, at present, talk to each other so there is a natural need for interconnection. Although commercial pressures have ensured that such links have not appeared in the past, in the face of the advantages of X.400 communications such arguments do not now hold water. Already far-sighted companies like GE IS are not only looking at X.400 but are also getting seriously involved in projects (see Chapter 7). Their work stems from a certain amount of user pressure as the gradual movement towards Open Systems Interconnection (OSI) begins. The eventual aim for GE IS is to offer host OSI facilities, including X.400, which will allow their users this additional option when connecting to services such as Quik-Comm. GE IS are also working with Transpac, in France, to establish a link between the worldwide Quik-Comm user base and the Atlas 400 message handling service. Links between the other national electronic mail services mentioned would seem to be a natural progression using X.400 as the basis. Links with BT's Gold 400 service, allowing international access, would also seem to be a possibility.

The establishment of further X.400 ADMD facilities within the UK is another strong possibility which is not precluded by the regulatory framework of this country. It is possible that one or more of the electronic mail suppliers will migrate to X.400 software for their systems and hence become message handling service providers (eg Telenet from the USA). Such services may be in direct competition with Gold 400 or may attempt to attack particular vertical markets such as EDI transfer of particular industry sectors.

All these developments discussed are possible and others which are not yet conceived, but the integration and interconnection of so many different services and systems will bring its own problems. The need for a central registration body for X.400 O/R names, and indeed any other such schemes which may come into conflict, is crucial. Without this kind of firm framework it is likely that the growth of the UK messaging infrastructure could be stunted. If action is left to this late stage it will almost certainly delay developments and do little to instill user confidence.

Generally, however, it seems that the impact of X.400 within the UK could be quite substantial. The potential for breaking down existing interconnection and integration barriers is there and it only requires the parties concerned to adopt the right approach.

5 Worldwide X.400 Message Handling System/Service Developments

The preceding chapters of this book have discussed in general and technical terms the potential of the X.400 recommendations. A major reason for the usefulness of this standard, which has been highlighted already, is that it has been internationally agreed and so can form a basis from which a worldwide messaging infrastructure can be built. In an attempt to put this potential in perspective it is the intention in this chapter to give a flavour of some of the recent X.400 developments around the world. In this way it is possible to gain an insight into the level of acceptance and implementation of these recommendations, worldwide, and to see that the seeds have already been sown which will lead to global X.400 interconnection.

Recent developments would now appear to confirm the early indications that X.400 is on the verge of achieving worldwide adoption and use. All the major service providers and telephone authorities now seem to be announcing that they will be offering X.400 services in the near future (see CCITT X.400 Questionnaire results at the end of this chapter). Some, like NTT (Japan) and Telenet, are already offering services; others, such as the Swiss PTT, are currently (January 1988) involved in pilot trials which will lead to the provision of a full service. When such services are established connections to other compatible X.400 services around the globe will be the logical next step. The provision of gateways and access units to facsimile, telex, teletex and electronic mail services will also mean that with a connection to only one service—an X.400 service—users will be able to access almost every corner of the world.

INTERWORKING DEMONSTRATIONS

As we move towards global interconnection the implications of the recent interworking demonstrations will achieve far greater significance. These demonstrations, at least in the case of ceBIT and Telecom 87, have been effective trials of the potential world X.400 links between different implementations and services. This will serve only to ease the establishment of formalised links when they appear, because the technical difficulties have already been addressed. In the case of the service platform for Telecom 87 (nine services connected) this is still established and being used by the Swiss PTT in the trials of their arCom 400 service. This should not, however, in any way diminish the task of overcoming such issues as inter-domain (connections between and across services) charging agreements between the service providers. This issue alone is likely to have a substantial impact on the acceptance and usage of such services, and hopefully sense will prevail and ensure that such facilities are not too expensive to use.

The interworking demonstrations also serve at least two other useful purposes:

— they are useful dry run for the crucial area of OSI conformance and inter-operability testing. Because OSI standards are written in a normal language-based format they are open to ambiguous interpretation by the implementor. To ensure successful interworking between implementations they have to be checked to establish that they are conformant to the original specifications. Product vendors and service providers participating in the interworking have the opportunity to connect to other X.400 implementations well in advance of the user. In this way potential user problems should be eased and migration to X.400 will become substantially less fraught;

— the extremely important role in heightening the awareness of X.400 standards and implementations and their viability as a solution to today's interworking problems.

All of the points mentioned so far and others which will be considered in detail in this chapter seem to point the way to X.400 becoming the basis of most of the future messaging requirements, worldwide. Another important point to consider when looking at X.400 is its position as the first viable OSI application. In this context its success would seem to be essential as a means of increasing the user awareness and acceptance of OSI and data communication standardisation generally. Hopefully X.400 should be able to assume this responsibility easily, as its potential for added value over what exists is already immense.

The ceBIT 87 X.400 demonstration, in Hanover, was the first of its kind to place the emphasis on available X.400 solutions. Each of the participants in the project (initiated in autumn 1986), either product vendors or service providers, had to commit to the provision of X.400 products and services during 1987. In the past there had been other X.400 demonstrations, notably:

1984 – Kokusai Denshin Denwa and University of British Columbia, a transpacific link of independently developed systems;

1985 – The X.400 demonstration at the SICOB exhibition. This involved three companies Bull, ICL and Siemens;

1986 – July, public demonstration of nine systems, primarily Japanese, but including a link to Xerox (Sunnyvale in the US).

These, however, involved the use of prototype systems to show that such connections were possible utilising the X.400 recommendations. Such demonstrations, although useful in heightening the awareness of the standards and showing that interconnection was technically possible, did little to give any idea of when practical products would be available.

The 14 organisations involved in ceBIT 87 (see list in Figure 5.1) were drawn from a range of seven countries, so its claim to be an international event was well founded. The interconnection network for the event (see Figure 5.2) largely centred around the interconnection of the Deutsche Bundespost (DB) and British Telecom (BT), ie the two X.400 services. The rest of the participants, product vendors, were either connected to one or both of the X.400 services or to other X.400 products. An amusing aside from the demonstration was that under FRG law it is only possible to have one carrier on German soil, that being DB. Hence in order for the demonstration in Hanover to proceed it was necessary to declare the exhibition standard as neutral territory, thereby circumventing the rule!

As with any large demonstration of this kind, independent project management is essential to ensure progress is achieved and that issues can be settled by an independent third party. The UK based Level 7 Ltd's consultancy operation was selected to provide project management for the ceBIT X.400 demonstration. They were involved in all aspects of the co-ordination and organisation for the event, including promotion and interworking tests.

In order to make the event as successful as possible the promotional aspects were taken quite seriously. A special X.400 logo was designed which appeared on free badges and brochures, given to people who attended the stand. There were 25,000 copies of this brochure distributed prior to the event and a further

British Telecom	Bull	Data General	Post
Digital	Hewlett Packard	ICL	NTT
Nixdorf Computer	Olivetti	Philips	Siemens
	Sydney	Xerox	

Figure 5.1   List of the 14 Organisations Involved in ceBIT 87

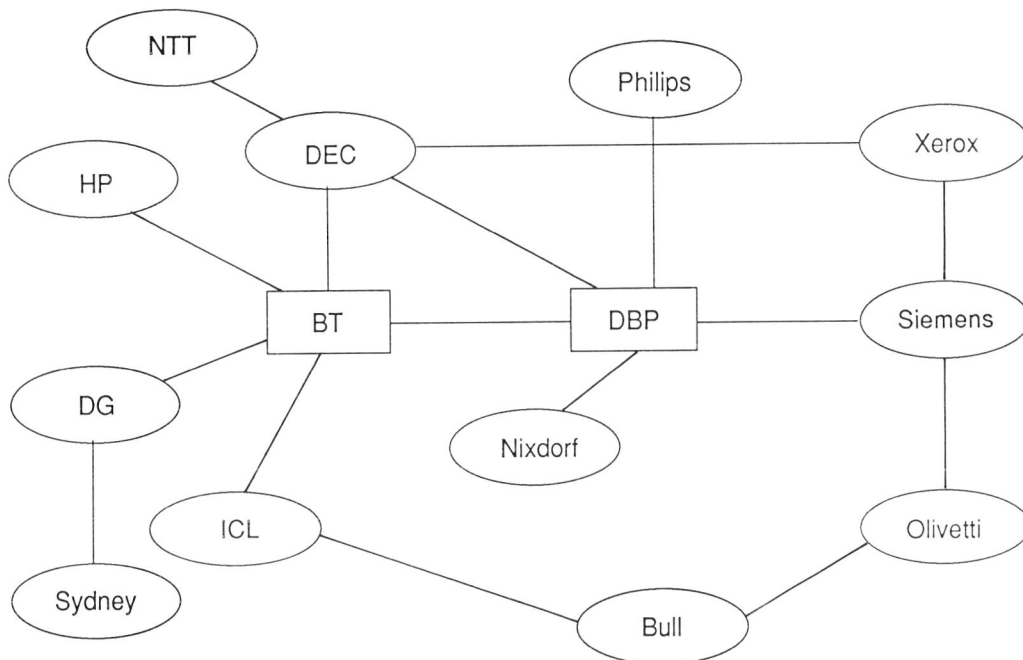

Figure 5.2   The ceBIT X.400 Demonstration

20,000 copies given away from the stand. This alone gives some indication of the impact that this well organised event had.

The interconnection/interworking for the demonstration was based around two European functional standards from CEN/CENELEC and CEPT:

— ENV 41201 – for connection between the PRMDs;

— ENV 41202 – for connection to and between the ADMDs, ie BT and DB.

The testing for the event was undertaken by SPAG Services (see Appendix 6) with all the participants being connected at some stage.

The demonstration for the event was of purely text messaging, and to ensure that it would give a useful insight to X.400 at work a test scenario was devised. This was based around a representative business application, the call for and receipt of tenders for a job, accomplished via the use of electronic messaging rather than the more normal postal means. These demonstrations were provided in a total of five languages and by the end of the exhibition more than a thousand demonstrations had been given; yet another indication of the impact of this event.

TELECOM 87—THE MAJOR X.400 EVENT OF 1987

Telecom 87 is a Trade Fair organised by the International Telecommunications Union (ITU), and it occurs every four years. It is regarded as a major event in the international telecommunications world, and attracts many hundreds of thousands of visitors. This event hosted the second major X.400 demonstration of 1987, and arguably the most significant.

The first obvious attribute of the Telecom 87 X.400 demonstration was its size, the largest of its type so far, with a total of 21 organisations involved (see Figure 5.3). This was by no means its most important attribute, however, because this event gave an insight into the future global X.400 messaging infrastructure. Nine of its participants were Telecommunication Administrations or Service Providers who will provide backbone X.400 Administration Domains (ADMDs) within their various countries of origin. These nine X.400 services were interconnected to form the basic platform for the demonstration, which will bear some relationship to the actual 'live service' links shortly to be established.

Previous X.400 events, including ceBIT 87, have had a relatively small presence of service providers. A great deal of interworking tended to be carried out between private domains: this is fine for achieving fixed connections between individual vendor's equipment, but the real power of X.400 messaging is unleashed when it is supported by major administrations. These X.400 services will offer focal points within their countries for connectivity, provision of international gateways to other countries, and access to the 'critical mass' of users so sought after by electronic mail suppliers.

With such a large and complex demonstration, careful project management was required to ensure its success, and with their experience in this area the UK company Level 7 Ltd were again selected. Interworking for the project was based mainly around ENV 41201 and ENV 41202, although some systems conformed to North American NIST and Japanese JUST profiles. Conformance was checked via interoperability testing.

The success of this demonstration as a means of increasing X.400 awareness can be judged by some of the statistics from the event:

Brochures

— well received;

— 60,000 printed;

— 15,000 distributed at Telecom 87;

— extensive distribution prior to the event and continuing after.

Traffic through the stand

— 11,500 people through;

— 800 sales leads obtained.

All in all it is a graphic indication of how an event of this kind can raise user awareness of X.400 as a viable solution.

EUROSINET

EurOSInet was conceived in January 1986 with just five suppliers (DEC, ICL, Hewlett Packard, Honeywell and Intel) involved. It was officially launched by John Butcher, then the Under-secretary at the Department of Trade and Industry, in June 1986. The stated objective is,

> To raise the level of market awareness and confidence in OSI as a practical solution to problems of systems interworking by offering a continually available demonstration of interworking between suppliers' systems using agreed OSI application profiles and, where appropriate, emerging standards.

It is a pre-competitive collaboration with well-defined code of practice which forbids public comment unless sanctioned by the Steering Group. Given the difficulties created by having a group of competitors in collaboration, this is not an unreasonable stance. Since the early days membership of the group has grown and now includes:

AT&T	British Telecom	Danet	Philips
Alcatel	PTT	Post	Dialcom Inc
Digital	Sydney	Telenet	Hewlett Packard
IBM	KDD	Telesystemes	Telic Alcatel
Transpac	NTT	Nixdorf Computer	Olivetti
		Unisys	

Figure 5.3   Names of the 21 Organisations in the Telecom 87 X.400 demonstration

— British Telecom;

— Concurrent Computer Corp Ltd;

— Cray Research (UK) Ltd;

— Data General Ltd;

— IBM;

— Logica;

— Tandem Computers;

— UNISYS Corporation;

— Wang UK Ltd.

The first EurOSInet demonstration was provided at the launch of the project by a subset of the original membership group, this was of OSI FTAM. Since then the demonstrator has undergone development and been shown at other events, notably the 1987 NCC National Technology Conference, and EurOSInet has developed a joint OSI—X.400 and FTAM demonstrator, first shown at the 1987 Compec exhibition in London. In order to broaden the scope of the project, EurOSInet have formed liaisons with OSI groups in the United States (OSINET) and Australis (OSICOM). This demonstrator is now available on a 24 hours-a-day basis and can be accessed around the world via X.25 based data networks.

Figure 5.4 shows the present configuration of the X.400 and FTAM demonstrator, as shown at Compec 87. Not all the participants were demonstrating both applications and this is indicated by the list below:

	X.400	FTAM
AT & T (US)	*	*
British Telecom (UK)	*	
Data General (UK and AUS)	*	
DEC (UK, AUS and CAN)	*	*
ENVOY 100 (CAN)	*	
Hewlett Packard (UK)	*	
Honeywell Bull (UK and US)	*	*
IBM (US, FR and AUS)	*	
ICL (UK)	*	*
ITL (UK)	*	
Intel (UK)	*	
Overseas Telecommunications Commission (OTC) (AUS)	?	?
Wang (UK)	*	

THE ESPRIT PROGRAMME

The European Strategic Programme for Research and Development in Information Technology (Esprit) was launched by the European Commission in Fenruary 1984. The programme was established as a result of initiatives taken by the European Commission and the so-called Round Table of 12 leading European IT firms. These initiatives stemmed from a growing concern at the European IT industry's poor competitiveness.

The stated objective of the Esprit programme is to encourage collaboration between European IT organisations in pre-competitive IT research and development. It is hoped that this collaboration will extend to the marketplace, to help strengthen the European technology base, and to help Europe become competitive in world IT markets.

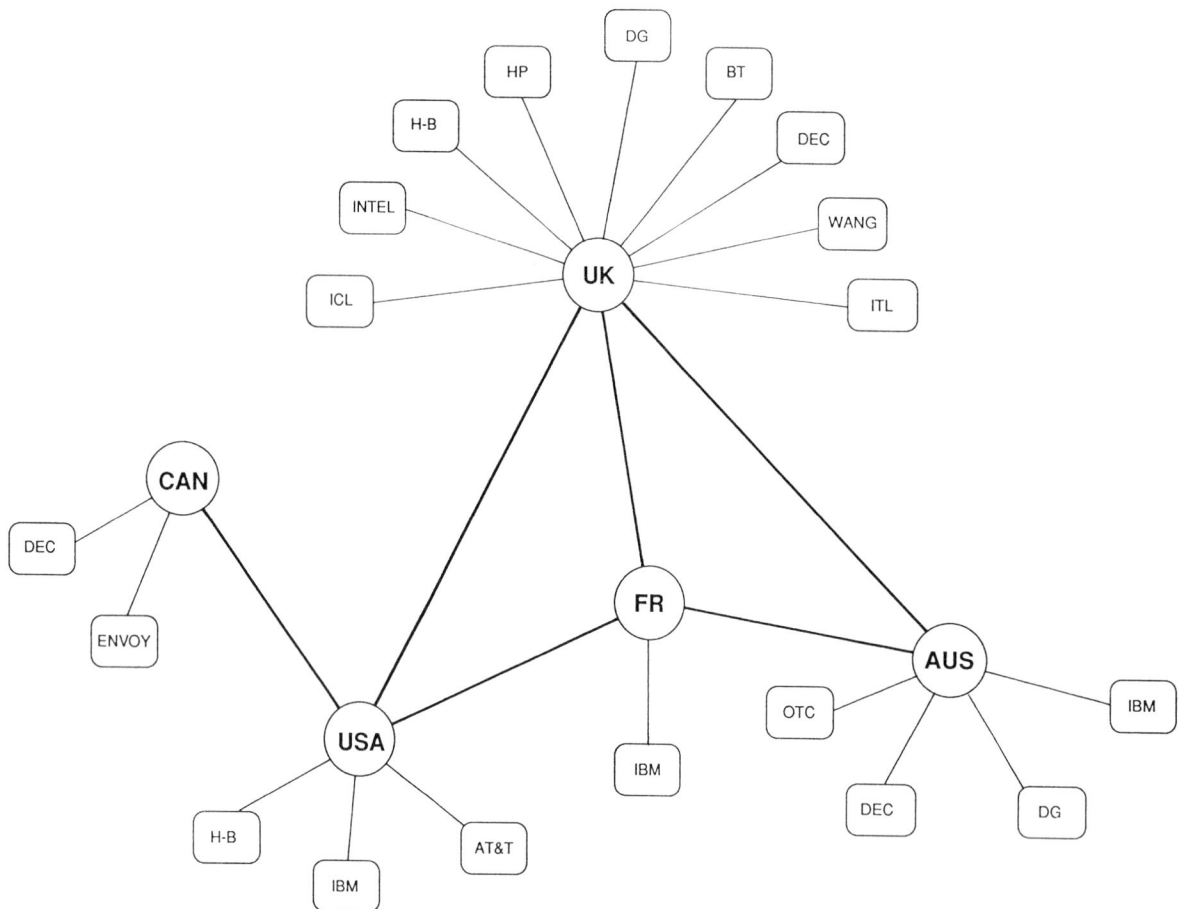

Figure 5.4   Configuration of X.400 and FTAM Demonstration

Esprit was initially set to run for five years with a budget of 1500 MECU, half of which is provided by the European Community and half by participating organisations, through shared cost contracts.

The technical areas encompassed by the Esprit programme include:

— Microelectronics (MIC);

— Software Technology;

— Advanced Information Processing;

— Office Systems;

— Computer Integrated Manufacture.

Because of the size of the project and the difficulty in achieving effective information co-ordination, distribution and exchange, it was decided that a special system would be required. The Esprit Information Exchange System (IES), a project in itself within the Esprit programme, is being established to enable the Esprit participants to exchange information effectively.

A decision was taken early in the development of the IES that in order to allow equipment from different vendors to be connected, the use of OSI standards was essential. Hence the general policy in the evolution stages of the system is to provide support for the development of interworking products based on OSI standards which will act as its basis in the future. Meanwhile, an immediate requirement for services will entail the adoption of proprietary solutions.

The first service to be provided to the IT R&D community under the IES programme was electronic mail and conferencing, via a system called EUROKOM initiated in August 1983. After four years the system is now used by over 50 per cent of the Esprit community. One of the future developments of this system, being undertaken by the University College of Dublin, is the provision of an X.400 mail gateway linking EUROKOM initially to the European EAN (academic community X.400 system) network and eventually to other X.400 sites. This gateway has already been successfully tested and plans for a 'live' service are well underway.

Research Open Systems for Europe (ROSE) is another Esprit IES project which is intended to develope OSI communications software for the UNIX portable operating system. There are four participants in this project, GEC, ICL, Olivetti and Siemens, and so far the work has included the development of software for both X.400 message handling and File Transfer, Access and Management (FTAM). Already a pilot scheme has been operated, to demonstrate interworking between different implementations, which consists of 20 nodes in five countries. It is anticipated that these developments (UNIX portable X.400 and FTAM software) will be exploited during 1988/89 with the participants producing commercially available implementations.

Finally, another interesting project within the Esprit IES programme is Communications Architecture For Layered Open Systems (CARLOS). The aim of the project is to develop generalised interfaces between systems conforming to OSI standards and those which do not, ie systems conforming to proprietary standards. The CASE communications company, along with Fischer and Lorenz, the University of Madrid and the University of Barcelona, have secured funding under this project to work in the area of X.400 product development. This specific work area is entitled Carlos Addition for Clustered Terminal User Agents (CACTUS) and is a project to implement X.400 in a form suited to medium-size private organisations. CACTUS will allow groups of people using simple terminals and personal computers to access X.400 facilities, with X.25 as the transport mechanism (see Figure 5.5). In order to achieve this, CACTUS will support P1 (MT service), P2 (Interpersonal Messaging) and the new P3+ mailbox client server protocol (see Appendix 7 and P7 in Chapter 10).

SERVICE PROVIDERS AND PTTs

This section is intended to give an insight into the type of X.400 developments which are being carried out globally by the providers of electronic mail services and the Postal, Telegraph and Telecommunications Authorities (PTTs). NTT, Telenet, Geisco, Deutsche Bundespost and the Swiss PTT have been selected for detailed comment because their respective approaches are interesting. The results of the CCITT X.400 MHS questionnaire (final section of this chapter) give a true insight into X.400 activities on a worldwide basis.

NTT–Personal Computer Communications Network

NTT's Personal Computer Communications Network (PCN) was the first of its kind to offer electronic mail, bulletin board, real time database access and a variety of other data communications services while conforming to the X.400 recommendations. After a year's trial the NTT-PC Communications Company, a subsidiary of NTT, began to offer a commercial service in November 1986.

To allow connection of personal computers to the network, NTT-PC supplies an adaptor and software conforming to OSI X.400. This allows transmission at 4800 bps over ordinary telephone lines with a high degree of reliability. Running NTT-PC's JUST-PIA software on the personal computer then makes it possible to achieve dial-up access to the NTT-PC Network.

The NTT-PC Network (see Figure 5.6), allows the user to exchange messages with other PC users and to gain access to other services and facilities, such as:

— other X.400 systems, when connections are established;

— telex and teletex;

— the Captain Videotex Network;

— bulletin board;

— database access.

In Japan the Ministry of Posts and Telecommunications have produced their own OSI profile of layers

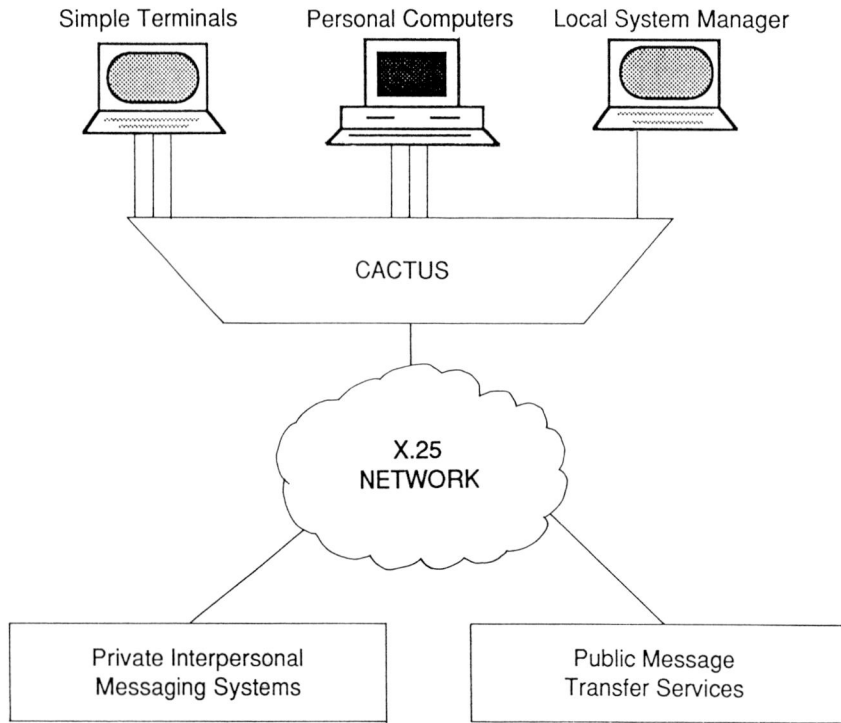

Figure 5.5   CACTUS from CASE Communications

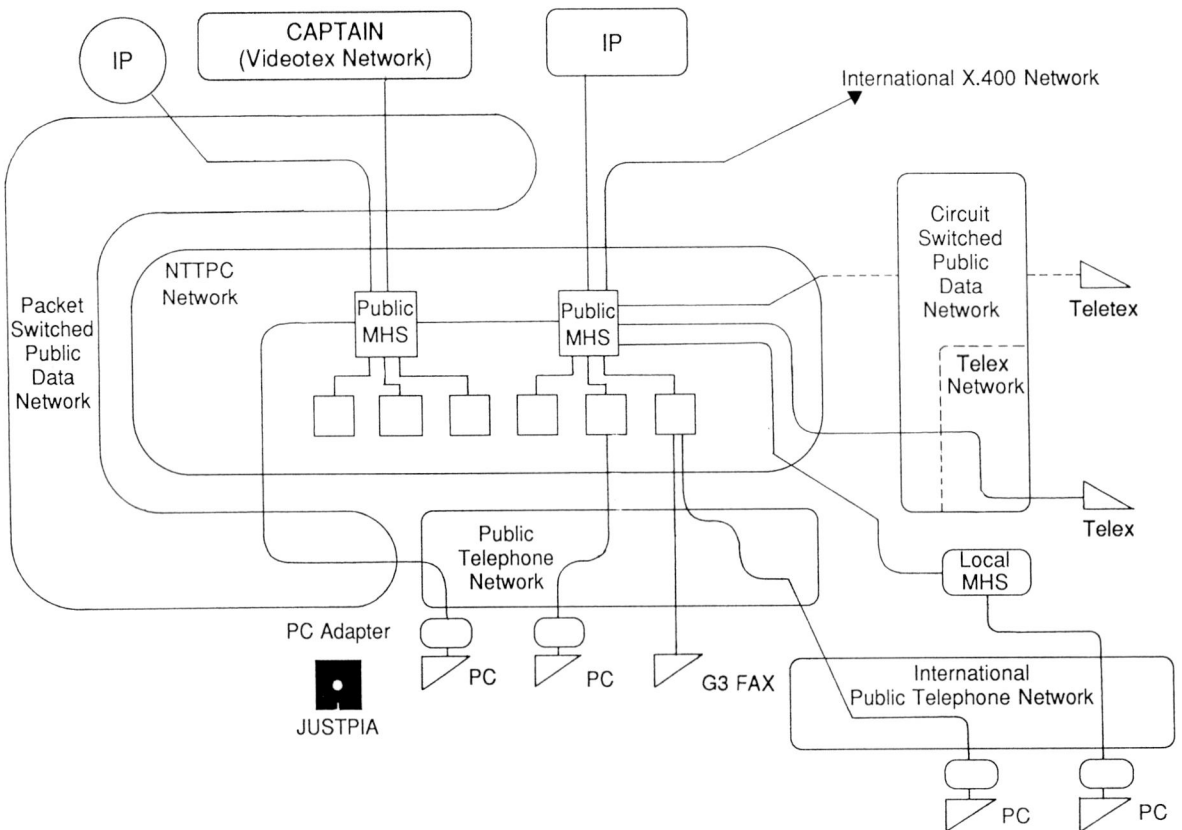

Figure 5.6   NNT–PC Network

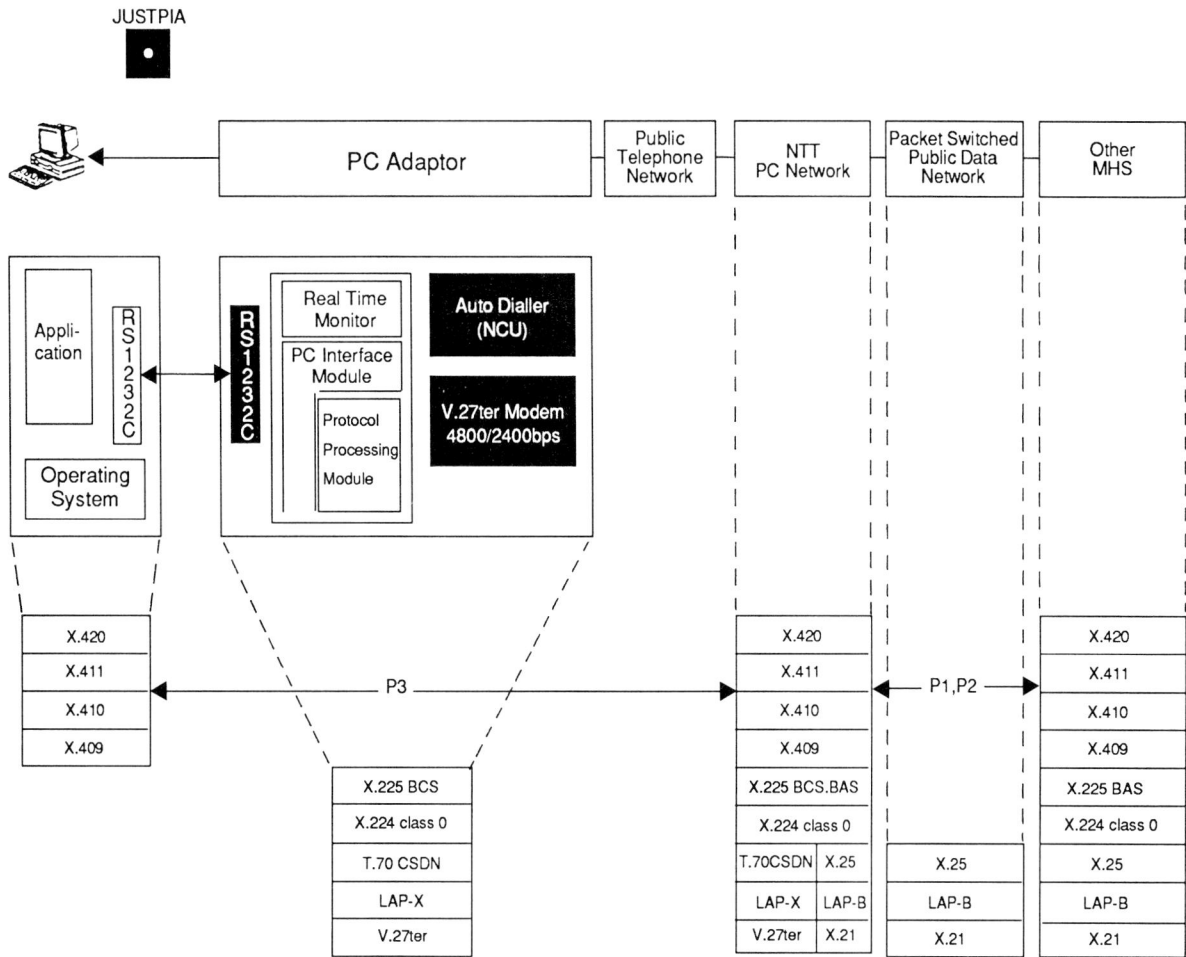

Figure 5.7 NNT–PC Adaptor and Connection to the PCN

1 to 5, specifically for personal computers. The Japanese Unified Standards for Telecommunications-Personal Computers, or JUST-PC for short, acts as the basis for the JUST-Message Handling System (JUST-MHS) work which profiles the CCITT X.400 MHS recommendations.

NTT's PCN is based on both JUST-MHS and JUST-PC, as indicated in Figure 5.7. The JUST-PIA software for the PC is performing the X.400 part of the operation utilising the improved version of the P3 protocol, P3+, which is included in the 1988 MHS recommendations (see Chapter 10), and which uses the concept of a simple terminal (the PC) communicating with a remote message store (situated at the PCN). The PC Adaptor, which is linked to the PC via an RS232C link, performs the functions associated with the lower five layers of the OSI model. An Auto Dialler and V.27 modem are included within the Adaptor to enable it to utilise the public telephone network to gain connection to the PCN.

NTT's PCN provides an excellent example of the flexibility of the X.400 recommendations: they not only cater for the connection of powerful computer-based message systems but also address the other extreme of simple terminal-to-terminal communications and message store access. In summary then the features of NTT's X.400-based PNC are as follows:

— interoperability with other message handling systems/services and with many types of personal computers;

— multi-media capabilities which allow the transmission of not only text messages but also images, programs and data files;

— message delivery to facsimile, telex and teletex;

— connection with a variety of other networks, such as videotex;

— fast access (4800 bps) and reliable transmissions due to the incorporation of a V.27 ter modem in the Adaptor and the use of High Level Data Link (HDLC) protocols.

Telenet X.400 Gateway

The Telenet Communications Corporation, in the United States, offers an integrated family of global electronic messaging and document transfer products, systems, and services. These include:

— Telemail electronic messaging service;

— Telemail Telex integrated connection to worldwide telex networks;

— Telemail 400, global messaging service;

— Telemail Xpress, hard copy delivery of laser printed letters;

— PC Telemail, communications software for personal computers.

All of these services and products reside under the umbrella name of Telemail Plus. There are more than 200,000 Telemail electronic mail service users worldwide and Telenet have furnished Telemail electronic mail service-based systems to 20 PTTs and other organisations throughout the world.

In the first half of 1987 Telenet established the X.400 gateway service, Telemail 400, for their Telemail electronic messaging service. This will allow Telemail users to exchange messages with other X.400 compatible systems and services when connections have been established. Currently, the service is available between the United States and the Envoy 100 electronic mail service in Canada and to ACEMAIL electronic mail systems in Japan. This allows subscribers in Canada and Japan to send and receive messages directly to Telemail electronic mail service subscribers in the United States. Other tests have also been carried out in connecting to integrated office systems such as Data General's CEO and Digital Equipment's All-In-One.

In addition to their X.400 gateway facility, Telenet have also perceived the requirement for X.400 conformance testing (see Appendix 6) in order to fill the breach until the arrival of the formal Corporation for Open Systems (COS—a joint NCC project) X.400 test centre in the US. Telenet have had some experience in this area as they have offered X.25 conformance testing in the US for almost a decade now.

Telenet's X.400 gateway operation would seem to point the way for many providers of electronic messaging services. By integrating X.400 into their existing system they have opened up a doorway which will eventually allow their users to be connected to all parts of the globe. This approach would now seem to be the only sensible solution for the providers of existing electronic mail services.

General Electric Information Services (GE IS)

It is more than 20 years ago now that the American company General Electric established its information services arm called the General Electric Information Services Company (Geisco). Today, GE Information Services Ltd, as it is now called, with its international headquarters in the UK, has built the world's largest commercially available teleprocessing network and also offers specialist software, business consultancy and systems development services.

GE Information Services (GE IS) have implemented many telecommunication-based computing systems and services with application areas as diverse as international systems for banking, financial markets, distribution management and international trade. The services GE IS offer are based on their three large interconnected computer 'super-centres' which can be accessed from more than 60 countries. In most instances this is possible by access to in country GE nodes either directly or via public data networks.

QUIK-COMM is the GE IS electronic messaging and information retrieval service that can be accessed through GE IS software available for their clients' systems. Users of this service have facilities such as sending messages or documents into other people's mailboxes, retrieving their own incoming mail, bulletin boards, look-up directory and mailing lists. GE IS also offer global EDI services based on their EDI*Express Service. In the UK these services are primarily offered via a joint venture company INS, which focuses upon a number of vertical markets including retail, insurance, motor, health, etc.

GE IS have been quick to realise the need for services based on the principles of Open Systems

Interconnection (OSI - see Appendix 4) both from their involement in the standardisation process and the requirements of clients. The concerns of their clients largely revolved around two areas:

— conformance to International Standards - increasingly clients have shown a tendency to produce requirement statements which make reference to OSI standards;

— 'Application Bridges' out of favour - negative views were being expressed on the use of 'application bridges' between mail systems; preferable to use recognised international standards instead.

In response to this, GE IS have started a number of significant development projects which will lead to their clients eventually having the option of communicating via OSI standards, including X.400. The remainder of this section considers the GE IS X.400 link to their QUIK-COMM service.

QUIK-COMM To X.400 Access Service

The X.400 Access Service product is designed to support message transfers between a QUIK-COMM system and worldwide public and private message handling services which support the CCITT X.400 recommendation. Entrance to X.400 services is via the Mark 400 Message Transfer Agent (MTA) which is provided within the GE IS Mark III host system. An Access Unit (AU) based on the QUIK-COMM connector, formats messages and acknowledgements for sending and receiving and is used to interface QUIK-COMM system users and Mark 400 MTA.

How Does the X.400 Access Service Operate?

The X.400 access service provides the following:

— QUIK-COMM system to X.400 messages;

— X.400 to QUIK-COMM system messages;

— X.400 compatibility with existing connector products.

These aspects of the service are described below.

QUIK-COMM System to X.400

The X.400 interface is based on the standard QUIK-COMM connector formats. Setting up an interface requires an X.400 connector system name and a corresponding community name assigned for the AU. There are two addressing methods which are valid for the X.400 address field. The address field can contain either a Message Handling System (MHS) Directory subscriber name or an X.400 set of Originator/Recipient (O/R) name attributes. In addressing the use of an MHS Directory Service, the Mark III MTA is designed to translate a subscriber name into a valid X.400 O/R address.

The X.400 message consists of an Interpersonal Message (IPM) header and body part. Not all of the IPM service elements are available for a QUIK-COMM system user to specify. The X.400 Access Service uses a default value for those mandatory fields and omitted optional fields which cannot be specified by QUIK-COMM system users. The X.400 body part contains the text of the QUIK-COMM system item.

All messages sent from a QUIK-COMM system via the Mark III MTA are sent with a confirmation request. This causes the final MTA responsible for delivery of any address to return a delivery report. This delivery report is returned by the final MTA when the message is successfully passed to the User Agent. If the message is undeliverable by the final MTA, a negative delivery report is returned.

The processes involved in message transfer from a QUIK-COMM system to X.400 are as follows:

— Outbound Message Notification;

— Outbound Message Connection;

— Outbound Message Transfer;

— Outbound Message Recording;

— Outbound Message Acknowledgements.

X.400 to a QUIK-COMM System

The Mark 400 MTA is set up with the country name, the Administration Management Domain (ADMD), and the optional Private Management Domain (PRMD). The organisation name is used to identify the desired Mark 400 Service (ie QUIK-COMM v EDI*Express). The use of the community name for the QUIK-COMM system organisational unit name (ie 'GE IS') is used to assure uniqueness among all QUIK-COMM systems. This community name is also used as the owner name for MHS directory access.

In receiving messages from an X.400 system, the X.400 Access Service searches the MHS Directory for the originator's O/R name. If a match is found, the corresponding subscriber name is entered as the *From* address. If there is no match, the *From* address would be:

O/R Name @ X.400 £

The O/R name would be a string of address attributes separated by attribute field delimiters. For a quick summary of these attributes, refer to the QUIK-COMM to X.400 Addressing Quick Reference Guide. The processes involved in message transfer from X.400 to a QUIK-COMM system are as follows:

— User Agent Delivery Notification;

— Inbound Message Connection;

— Inbound Message Transfer;

— Inbound Message Acknowledgement.

X.400 and Existing Connector Products

There are no restrictions for existing QUIK-COMM connector systems (ie DEC ALL-IN-1, IBM DISOSS and PROFS) to send and receive messages through the X.400 Access Service. The interface will operate the same as any other connector-to-connector exchange.

The ability to address an X.400 O/R address will be based on the individual connector addressing restrictions. Through the use of QUIK-MHS Directory all connector systems should be able to address X.400 systems. X.400 users wishing to send messages to QUIK-COMM connector users will need to know the community name of the QUIK-COMM system to which the External Mail system is attached.

THE CCITT QUESTIONNAIRE RESULTS

Question: 13/I (Message Handling Service)

STUDY GROUP I — Contribution 134

SOURCE: Q13/I Rapporteur (Mr McKnight)

TITLE: Results of MHS Questionnaire

---

1. Introduction

The MHS Questionnaire created by Study Group I was sent to all members of the CCITT as CCITT Circular No. 52 dated January 5, 1987. The intent was to determine the plans of Administrations and RPOAs with respect to potential interconnection of X.400 based message handling services. The responses received as of May 13, 1987 are summarised in Annex 1.

2. Explanation of Summary

The results in Annex 1 are shown only for participants that responded positively to Question# 1. Names of Administrations and RPOAs that responded stating that they were not planning to implement X.400 MHS are listed in Annex 3. If replies continue to arrive they will be summarized in Study Group I working documents only. Columns left blank signify that respondent either left it blank, or made a comment such as "not decided yet", or "under study".

3. Contact Person

The attached annex 2 lists the contact names that were supplied on each positive response.

4. Legend for Annex 1

Question 6   I - Input, O - output, and I/O - input and output.

Question 8   EDI - Electronic Data interchange also called EBOI.

Annex 1 — Q13/1 MHS Questionnaire Results

Q	Question (Respondent →)	Australia Telecom	Austria Radio	Belgium RTT	Brazil Telerj	Canada Telecom	Canada CNCPT	Chile Telex-Chile	Denmark Telecom	Deutsche Bunderpost	Finland PTT
1	Planning to Provide X.400 Svce.	YES	YES	YES	YES	YES	YES	YES	YES	YES	YES
2	Providing EMS Today	YES	YES	YES	YES	YES	YES	YES	YES	YES	YES
3.1	Planning to Provide IPMS	YES	YES	YES	YES	YES	YES	YES	YES	YES	YES
3.2	Pilot/Trial Start Date		1988	1987	1989	1987	1987			1987	1987/88
3.3	International Availability Date		1989	1987		1987	1987		1988	1987	1988
3.4	Access from Following:										
	IA5		YES	YES	YES	YES	YES	YES	YES	YES	YES
	P3 Type		YES			YES	YES	YES	YES		
	Telex		YES		YES	YES	YES	YES			YES
	Teletex				YES		YES				
	Facsimile				YES		YES				YES
	Videotex		YES		YES						
	Voice								YES	YES	
	Others				YES	YES					
4.1	Planning to Provide X.400 MTS	YES	YES	YES	YES	YES	YES		YES	YES	YES
4.2	Independently of IPMS			YES	NO	YES	YES		YES	YES	YES
4.3	Pilot/Start Date			1987	1989	1987	1987		1988	1987	1987/88
4.4	International Availability Date			1988		1987	1987			1987	1988
5.1	Interconnection to PRMDs	YES	YES	YES	YES	YES	YES	YES	YES	YES	YES
5.2	Interconnection to PDS		YES	YES	YES	YES	YES	YES		YES	
6	Non Subscriber Access In/Out		I/O	I/O		O					
	Telex Service	I/O		I/O	I/O		I/O	I/O	I/O	I/O	I/O
	Teletex Service				I/O		I/O		I/O	I/O	
	Facsimile Service	O	O		I/O		I/O	I/O	O	I/O	O
	Videotex Service	I/O	I/O		I/O					I/O	
7	Name of Service	KEYLINK	TELEBOX	DCS MAIL		Envoy 100	Dialcom	Correo Electrónico	Directory	TELEBOX	TELEBOX
8	Other Applications to Develop	EDI	EDI	EDI		EDI			Doc. Trans.	EDI	EDI
	"								Mess Store		
	"								I/C Radio-Mobile		
	"										
	"										

Annex 1 - Q13/1 MHS Questionnaire Results

Q	Question	France PTT	Greece OTE SA	Hong Kong C & W	Iran PTT	Italy PTT	Japan KDD	Malaysia	Mali	Netherlands
1	Planning to Provide X.400 Svce.	YES	YES	YES	YES	YES	YES	YES	YES	YES
2	Providing EMS Today	YES	YES	YES	NO	NO	YES	NO	NO	YES
3.1	Planning to provide IPMS	YES		YES	YES	YES	YES	YES	YES	YES
3.2	Pilot/Trial Start Date	1987			1991	1988	1985	1988	1990	1987/88
3.3	International Availability Date	1987			1992		1986	1988	1990	1987/88
3.4	Access from Following:									
	IA5	YES		YES		YES	YES	YES		YES
	P3 Type			YES		YES				YES
	Telex	YES		YES	YES	YES	YES			YES
	Teletex	YES				YES		YES		YES
	Facsimile			YES	YES	YES	YES		YES	YES
	Videotex	YES			YES	YES	YES	YES		YES
	Voice									
	Others	YES								
4.1	Planning to Provide X.400 MTS	YES		YES	YES	YES	YES	YES	NO	YES
4.2	Independently of IPMS	YES		YES	NO		YES		NO	YES
4.3	Pilot/Start Date	1987			1991	1987	1985			1988
4.4	International Availability Date	1987			1992	1988	1988			1988
5.1	Interconnection to PRMDs	YES		YES	NO	YES	YES	NO	NO	YES
5.2	Interconnection to PDS	YES		YES	YES	YES		NO	NO	YES
6	Non Subscriber Access IN/OUT	O		I/O	I/O		O			I/O
	Telex Service	O			I/O					I/O
	Teletex Service	O								O
	Facsimile Service	O		O	I/O		O			I/O
	Videotex Service									
7	Name of Service	ATLAS 400					MESSAVIA			MEMOCOM
8	Other Applications to Develop	EDI		EDI			File Trans.			EDI
	"	File Transfer		UB2 S & F			Media-Conversion			
	"	Press					Language-Translation			

Annex 1 - Q13/1 MHS Questionnaire Results

Q	Question	Norway Admin.	Spain CTNE	Sweden Televerkat	Switzerland PTT	UK BTI	USA AT & T IS	USA ITT WC INC	USA MCI INT.
1	Planning to Provide X.400 Svce.	YES	YES	YES	YES	YES	YES	YES	YES
2	Providing EMS Today	NO	YES	YES	NO	YES	YES	YES	YES
3.1	Planning to Provide IPMS	YES	YES	YES	YES	YES	YES	YES	YES
3.2	Pilot/Trial Start Date	1987	1988	1987	1992	1987	1987		1987
3.3	International Availability Date	1987/88	1988	1989	1994	1987	1987		1988
3.4	Access from Following:								
	IA5	YES	YES	YES	YES	YES	YES	YES	YES
	P3 Type	YES	YES	YES	YES	YES	YES	YES	Maybe
	Telex	YES	YES		YES	YES		YES	YES
	Teletex	YES	YES		YES	YES			
	Facsimile		YES		YES	YES		YES	YES
	Videotex		YES		YES	YES			
	Voice	YES	YES	YES	YES		YES		Maybe
	Others	YES	YES	YES	YES		YES		
4.1	Planning to Provide X.400 MTS	YES	YES	YES	YES	YES	YES	YES	YES
4.2	Independently of IPMS	YES	NO	YES	YES	YES	YES	YES	YES
4.3	Pilot/Start Date	1987	1988	1988	1992	1987	1987	1987	1987
4.4	International Availability Date	1987/88	1988	1989	1993	1987	1987	1988	1988
5.1	Interconnection to PRMDs	YES	YES	YES	NO	YES	YES	YES	YES
5.2	Interconnection to PDS		YES	YES	YES	YES	YES	YES	YES
6	Non Subscriber Access IN/OUT								
	Telex Service	I/O	I/O	I/O	I/O	I/O	I/O	I/O	I/O
	Teletex Service	I/O	I/O	I/O	I/O	I/O			I/O
	Facsimile Service	I/O	O	O	O	O		O	O
	Videotex Service	I/O	I/O	I/O	I/O	I/O		I/O	
7	Name of Service	TELE-MESSAGE	PWH & TS				AT & T MAIL		
8	Other Applications to Develop	EDI	EDI	EDI	EDI	EDI	EDI	EDI	EDI
	" " "		Directory		Telegram	EFT	Directory	File Trans.	
	" " "		Security						

Annex 2 – Q13/I MHS Questionnaire Contacts

COUNTRY	ORGANISATION	CONTACT	ADDRESS	COMMUNICATIONS
Australia	Telecom Australia	Jan Bylstra	7/199 William St Melbourne Vic. 3000	Tel +61 3 606 8653 Tlx aa 134875
Austria	Radio - Austria	Friedrich Pexa Tech. Director	Renngasse 14 A 1010 Wien	Fax +43 222 533 00 00 Tlx 114731 aa
Belgium	RTT	Mr Qualria Ingénier	Blvd Emile Jacqmain 166 12ème étage 1210 Bruxelles	Tlx 29257
Brazil	Administration	L.P.M.	Telerj Av. Presidente Vargas 2560 - 6 Andar Centro Rio De Janeiro	Fax 233-3757 Tlx 21795 TLRJ BR
Canada	Telecom Canada	John R Hunt	Room 1950 160 160 Elgin St Ottawa, Ontario K1G 3J4	Tel +1 613 567 0801 Tlx 053 3763 Fax +1 613 567 1453
Canada	CNCP Telecommunications	Mr O Stublis	3300 Bloor St West Toronto Ontario M8X 2W9	Tlx +21 6218362
Chile	Telex-Chile Comunicaciones Telegráficas S.A.	Sr. Ricardo Pedraza	Morandé No. 147 Santiago	Tlx 240330 Telex CL
Denmark	Telecom Denmark	Mrs A. M. Majgaard	Traffic Division Telegade 2 2820 Taastrup	Tlx 22999 Fax +45 2 529331
Federal Republic of Germany	FTZ	W Tietz Referal T 21	Postfach 50 00 D-6100 Darmstadt	Tlx 419511 Fax +496151 8346 39

Annex 2 – Q13/I MHS Questionnaire Contacts

Country	Organisation	Contact	Address	Communications
Finland	PTT	Ms Leena Save	TEO/lmt P.O. Box 526 SF 00101 Helsinki 122151	Fax +3580 704 2091
France	PTT Telecom France	J Guesnier	DGT/DAK 6 René Cassin B P 1833 35018 Rennes Cedex	Tlx 950 261 Fax +33 99 38 49 61
Greece	O.T.E.S.A.	Telex Data Transmission Division	5 Stadiou Str. GR 10562 Athens	Tlx 215482 DSS GR Fax +301 3473299
Hong Kong	Cable & Wireless (HK) Ltd	M Coggin	22 Fenwick St New Mercury House Wanchai	Tlx 60758 CWMCE HX Fax 5 8612143
Iran	Ministry of PTT	A Roshanfelor Rad	Directorate General of Telecommunications Dr Shariati Ave P.O. Box 11365-931 16314 Tehran	Tlx 212444 PTT IR
Italy	PTT Administration	D. Stiberio	Ministero PT Direzione Centrale Servizi Telegrafica Viale Europe 190 00144 Roma	Tlx 616454

Annex 2 – Q13/I MHS Questionnaire Contacts

Country	Organisation	Contact	Address	Communications
Japan	KDD	Talec Harada	2-3-2 Nishi Shinjuku Shinjuku-ku Tokyo 163	Tlx J22500 KDD TOKYO Fax +81 3 347 7000
Malaysia	Syarikat Telecom Malaysia Berhad	Mr P Sritharan	Switching Division Telecoms Headquarters 6th Floor Jaian Raja Chulan 50200 Kuala Lumpur	Tlx MA 90067
Mali	Office des Postes et Telecommunications du	Mme Tracre Halima Ronate	Inspecteur Telecom Division Exploitation Commerciale Direction des Telecom Bamako	Tlx 933
Netherlands	Netherlands PTT	Willem van Keulen	P.O. Box 30 000 2500 GA The Hague	Tlx 31255 Fax +31 70432040
Norway	Norwegian Telecommunications Administration	Mr Sigmund Undheim Undheim	P.O. Box 8701 St Olavsplass N 0130 Oslo 1	Tlx 71203 Fax +47 2 48 8720
Spain	Administration	José Antonio Bogas Morillo	Dirección General de Correas y Telégrafos Subdirección General de Explotación Sección de Nuevos Servicios CCP Chamartin 28070 Madrid	Tlx 27728 GENTLE Fax 34 1 4294621

Annex 2 – Q13/I MHS Questionnaire Contacts

Country	Organisation	Contact	Address	Communications
Spain	CTNE	A Cal	Telefonica Secretaria Tecnica Gran Via 28 28013 Madrid	Tlx 27320 Fax +341 2311459
Sweden	Televerket	Lars Rune Larsson	FNs S 123 86 Farsta	Tlx 14970 gentel s Fax +46 8 648678
Switzerland	PTT	F Maurer	Section Commutation de Messages Viktoriastrasse 21 CH 3030 Berne	Tlx 911010 pit ch Fax +41 31 62 25 49
Thailand	PTT	Mr Kittin Udomkiat Director	Telecommunication Services Division CAT Laksi Bangkok 10002	Tlx 80006 CAT TH
United Kingdom	British Telecom International	Mr Neil Dunnet	The Holborn Centre 120 Holborn London EC1N 2TE	Tlx 21601 BTI G Fax +44 1 831 9959
USA	AT & T Information Systems	Mr Paul Chu	307 Middletown Lincroft Road Lincroft New Jersey 07738	Tlx 156243351 PCHU Fax +1 201 576 7910
USA	ITT World Communications Inc.	Mr Peter Calistri	100 Plaza Drive Secaucus New Jersey 07096	Tlx 423301 Fax +1 201 330 5105
USA	MCI International	David Bland/ Karen McGowan	2 International Drive Ryebrook New York 10573	Tlx 6502757508

Annex 3

Administrations and RPOAs Not Planning to Implement X.400 at this Time

Macau – Administration – Direcção dos Serviços de Correlos e Telecomunicaçues de Macau

RPOA – Companhia de Telecomunicaçues de Macau, S.A.R.L. (CTM)

Egypt – National Telecommunications Organization - ARENTO

Cyprus – Cyprus Telecommunications Authority

Luxembourg – Administration des P. at T.

Columbia – Emprasa Nacional De Telecomunicaciones

Kuwait – Ministry of Communications (feasibility study under way)

6 Potential User Applications for X.400 Systems

In Chapter 3 the technicalities of the X.400 recommendations were discussed in simple terms, and it was stated that an intrinsic design feature of these standards allows systems based on them to act as all-purpose data transfer platforms. This feature is really the key to one of the major advantages of the X.400 approach in comparision with existing types of pure message-based electronic systems.

The architecture employed for X.400 systems includes a separation between the underlying message transfer mechanism and the actual user applications. In real terms this has been achieved by the separation of user services provided by the User Agents (UAs) and the message transfer (MT) service provided by the Message Transfer Agents (MTAs). At a protocol level this means that there is effective isolation between the user service protocols, 'P2 type' protocols, and the message transfer (MT) service and its associated P1 protocol. The MT service will provide the underlying means of message transfer while the user services will be concerned with the specific user applications. In this way the MT service is not concerned with the contents of a message and hence only transports it to the requested recipient. Because of this, X.400 systems can carry any type of structured or unstructured data with the interpretation of the information purely an activity associated with the application, ie the end-system.

The separation between the services allows the definition of new classes of user services, and hence User Agents (UAs), which will be established as and when a specific user need or application is perceived. Many applications will probably only become apparent when experience is gained with X.400 systems and people and groups start to consider the way to make the best possible use of their systems.

There are really two methods of creating a new user service—in the public domain or the private domain. In the public domain a new application would be defined as an addition to the original messaging standard with the same status as Interpersonal Messaging (IPM). In the private domain case an organisation or group will realise the need for a specific solution and then go ahead to propose an X.400-based solution in the guise of a new user service. It is desirable from a standardisation view point that the former of these approaches will be the most popular. In this way difficulties experienced attempting to interwork systems will be minimised. Although experience with X.400 implementations is still fairly limited, its potential as an application platform has already been realised. Various organisations have proposed specific applications and it is these which will be discussed in this chapter. Examples of these applications are:

— system/service integration and interconnection;

— interpersonal messaging;

— processable document interchange;

— electronic data interchange (EDI);

— a base for distributed systems.

These few examples, however, are probably only scratching the surface of the usefulness of X.400 systems and only actual user experience will be able to influence its evolution of the coming years.

Another potentially important area which will emerge is X.400 MHS 'home-grown' industries. This is the situation whereby organisations offer services which can be accessed via a connection to the public messaging environment. The potential in this area is at present very immature and will only become apparent as the messaging infrastructure develops, both within specific countries and around the globe.

Throughout the remainder of this chapter individual examples of some of the most likely applications for X.400 systems are discussed in more detail. Again it must be emphasised that these are just a small sample of the wide range of applications which can be operated over such systems, the full potential of which will only become apparent when experience is gained.

SYSTEM/SERVICE INTEGRATION

Some of the Problems

The electronic messaging environment which has emerged so far is one of divergent technologies. Systems have been developed and adopted by various groups with little consideration for their ability to interwork with the existing messaging infrastructure. Looking at the list of services and systems mentioned below, and as described within Chapter 2, this becomes readily apparent:

— telex;

— teletex;

— public electronic mail;

— private electronic mail;

— facsimile.

Surveying the world as a whole it is likely that one or more of these technologies would be in use in every business environment which could be located. This is particularly true of telex and facsimile, which seem now to be about equal in the popularity stakes, each with around 1.5 – 2 million users worldwide. Although this may be the case, worldwide electronic messaging is still not a reality without an individual connection to each technology. Even when this has been achieved, there is still the problem of ascertaining which system/service is being used by the intended recipient and which addressing scheme to be employed. This latter point has been a problem for the users of electronic messaging systems for some time now, particularly computer-based ones where the user has to deal with unmanageably long and unfriendly names/passwords.

The situation with private internal messaging systems is even worse. There are many integrated office systems and LAN-based electronic mail packages available and few of these have the ability to exchange messages. This leads to the development of gateways which are usually on a one-off basis and therefore expensive. In the past, different groups within organisations have been forced into the purchase of incompatible systems because of the specialised nature of their individual needs. Now, however, where sharing resources and systems integration is 'the order of the day' such groups are faced with a dilemma; whether to ignore the need for internal communications or to buy-in a solution – either a gateway or, in extreme cases, a complete new system.

External communications from an organisation also present a huge problem with links to multiple technologies having to be maintained in order to address all potential business associates. A situation which highlights the problems of intra/inter-organisational communications is the merger of two companies. Apart from the general logistics involved in this type of venture the consideration of efficient communications and information resource sharing becomes paramount. Because of the probability that each organisation is utilising different electronic messaging systems, a serious problem emerges. Effective communications in a large, merged company become crucial in the day-to-day running of such an organisation, hence the requirement for efficient electronic messaging. Integration of information resources is also a major consideration, and there is a need for electronic messaging as a means of transferring centrally held information throughout an organisation. Such problems need to be solved before the merged company can operate at its most efficient level.

X.400 – THE SOLUTION TO SYSTEM/SERVICE INTEGRATION

In view of the problems outlined so far in this chapter it is obvious that the most useful form of business communications tool would be one service/system which could gain access to all of today's messaging media. The internationally agreed X.400 recommendations have been defined to fulfil just such a role.

One of the underlying concepts behind X.400 was to attempt to define a service which could pull together all of the current messaging media so as to establish a global messaging infrastructure. In this way all business environments would be able to exchange messages electronically. In the past, the true benefits of electronic mail have been inaccessible because of the lack of interconnection and integration of services. X.400 overcomes such obstacles by allowing users to access all services from one terminal, using one standard addressing format, thereby allowing them to make the most productive use of electronic mail. But how will this be realised in practice?

The first things to consider are the X.400 services; these will be provided either by PTTs or service providers. They will act as the backbone carriers, providing a means of interconnection/access to other X.400 ADMDs and PRMDs (both national and international) and offer transparent gateway services to other messaging media, such as telex, teletex, etc.

For X.400 users who wish to connect directly to X.400 services the options are:

— X.400 conformant office systems;

— X.400 gateway from an existing office system;

— link directly into a public X.400 service.

It is also likely that users may not wish to interconnect to a service, but rather connect directly (via X.25) to another X.400 private system. This is entirely possible.

Users of existing messaging media will be able to receive mail from an X.400 user sent via one of the gateways provided by an X.400 service. When sending a message to an X.400 user this will be accomplished by the originator utilising a special X.400 address which would access a gateway which would take care of the address mappings, conversions, etc (eg BT Gold E-mail to Gold 400).

It is intended that X.400 systems will be used as the glue with which to bond together existing messaging services, as well as providing services in their own right. This means that these systems will both preserve and enhance existing technology investments. Consider the following specific examples as a means of illustration:

Example 6.1

A company has a large existing investment in a specific type of integrated office package which fulfils their internal electronic mail needs adequately. In such a situation an X.400 gateway package (which most major vendors now provide) would allow the internal mail system access to an X.400 messaging service such as BT's Gold 400 (see Figure 6.1). With this single additional connection the company would be able to send messages to all the other X.400 users connected to Gold 400 and, via the gateway services, to all the other existing electronic messaging media. In this way the company's investment in the office system is preserved (the only change is an additional package) and also enhanced because it can now access a wide range of other systems/services previously inaccessible.

Example 6.2

A company runs a large public electronic mail service (non-X.400) which offers a good service to its users, who can gain access to their mailboxes via a dial-up connection. The issue for such a company is whether it can provide a viable service in competition with X.400 systems. The solution to the problem is to establish a gateway between the public electronic mail service and a public X.400 message handling service (see Figure 6.2). In this way the facilities of the original service are maintained with the significant enhancement of a connection to the X.400 messaging infrastructure for the user.

Example 6.3

A company operates a telex-based communications system between its offices around the globe, which they deem to be totally adequate for their needs. In such as case the effect of X.400 would be purely

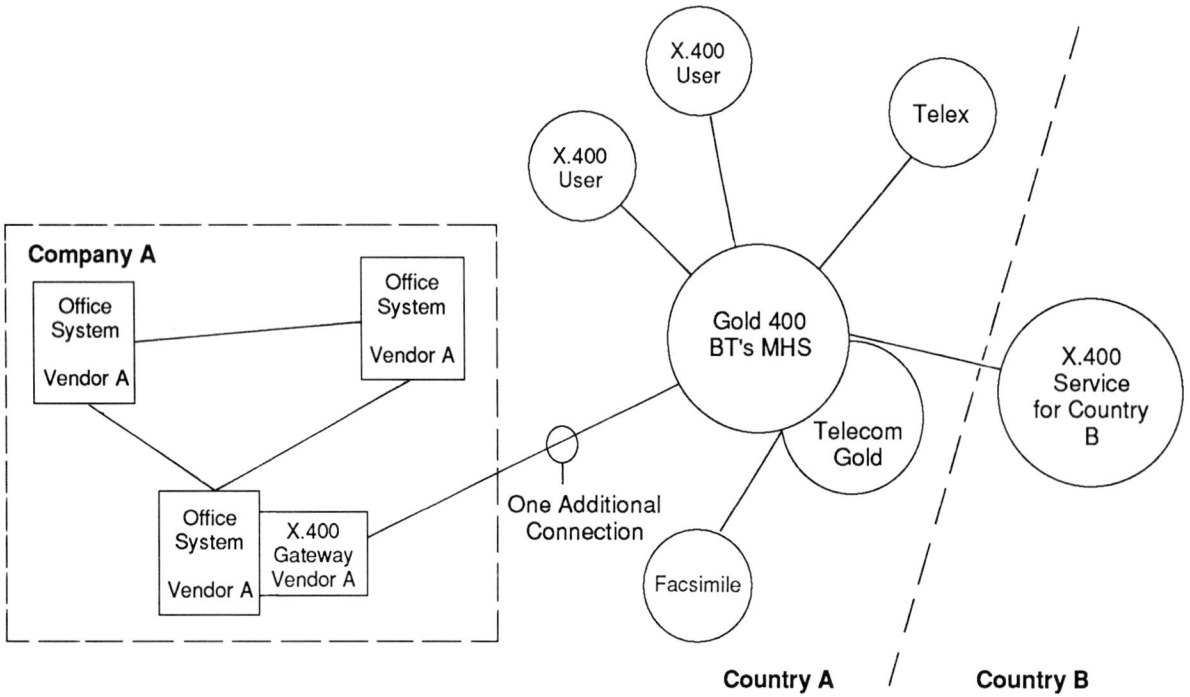

Figure 6.1   Illustration of Example 6.1

indirect, with the company involved in no extra expenditure. Because the X.400 services will provide two-way telex gateways the company will be able to take advantage of the X.400 connection while paying only for the gateway facility while it is used (see Figure 6.3). Again an existing system is enhanced, but in this case with minimal extra cost to the company.

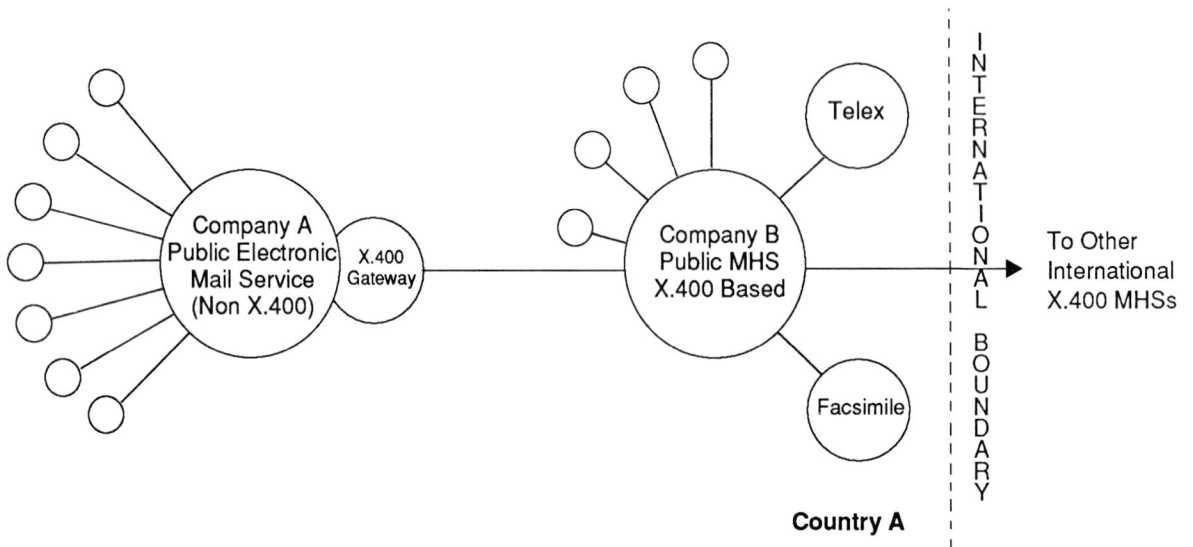

Figure 6.2   Illustration of Example 6.2

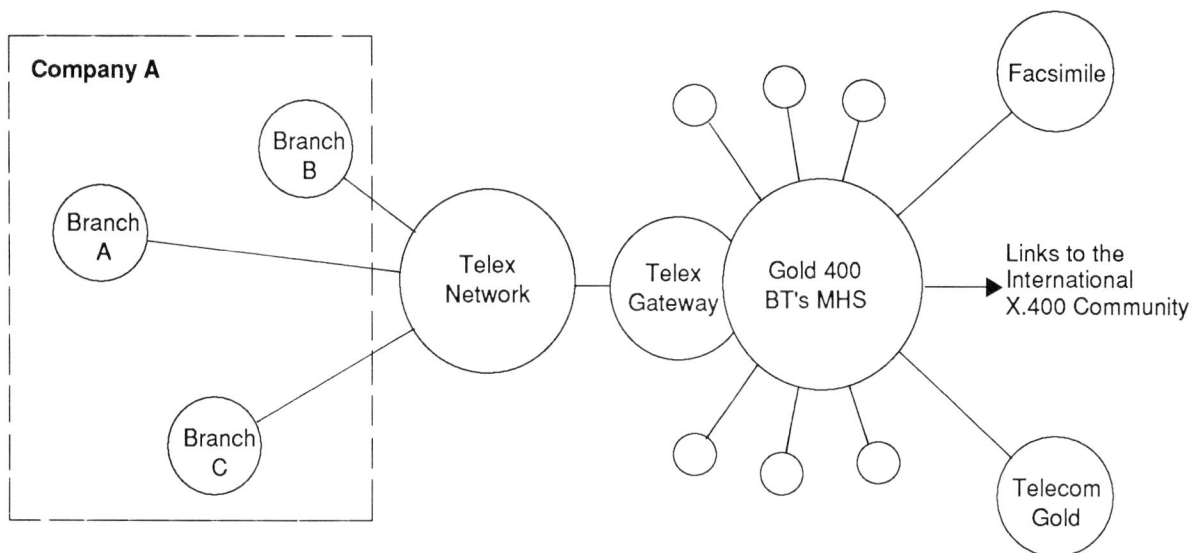

Figure 6.3   Illustration of Example 6.3

It should now be clear that the X.400 recommendations do not offer just 'another communications protocol'; its usefulness is substantially deeper than this. Because it is an internationally agreed solution it can act as a basis for all implementations, worldwide. In this way interworking between systems is effectively guaranteed, although this is not quite the case in practice (see later chapters for discussions). Any future development of these recommendations will continue in the theme of preserving existing investments and so will be backward compatible. X.400 systems will act as an extremely useful means of pulling together today's divergent messaging technologies to form a single coherent worldwide messaging infrastructure, allowing all electronic mail users to communicate. In this way X.400 should be the catalyst which will allow electronic messaging to come of age and prove its true worth to the user.

INTERPERSONAL MESSAGING

Interpersonal Messaging (IPM) is the only user service and class of UA to be defined within the original X.400 recommendations. The main reason for this choice is the existence of a ready market for such a service and therefore its popularity should be reasonably assured.

This service enables a human user to send IP-messages to one or more recipients. The service is provided by IPM User Agents (UAs) which communicate via the use of an IPM protocol known as P2. The additional elements of this service, over and above those of the Message Transfer (MT) service, are indicated by an IPM header which is attached to an IPM body part before it is placed in the P1 MT envelope. The service is generally similar to those provided by the public electronic mail services and includes the service elements listed in Table 6.1.

As well as sending mail to other IPM UAs and users it is also possible to send and receive messages from telematic terminals. Access to such devices is difficult because of their lack of sophistication, which means in some cases that they cannot communicate with computer-based hardware. To overcome this problem, access units – centrally held conversion software – will be provided by the PTT controlled Administration Management Domains (ADMDs). This will enable IPM users to communicate with users of telex and teletex on a send-and-receive basis. However, access to facsimile terminals will generally be on a message delivery basis only. This one-way access to fax is due to the problems in converting scanned images for display on text-based terminals. In the previous section it was stated that X.400 systems will act as a means of interconnection and integration of existing messaging systems and services, this is catered for by the provision of these gateway connections which will allow users of the IPM service to address other large user bases.

Blind Copy Recipient Indication

Non-Receipt Notification

Originator Indication

Primary and Copy Recipient's Indication

Expiry Date Indication

Importance Indication

Sensitivity Indication

Subject Indication

Reply Request Indication

Body Part Encryption

Table 6.1   IPM Service Elements

IPM UAs make use of the basic MT-service elements and also permit the optional facilities to be requested. In addition to these, the IPM UA will also provide the basic IPM service elements (see Table 6.1) which are concerned with uniquely identifying a message, the type of body part, and any specific attributes.

IPM UAs may also provide local editing and filing/retrieval capability in order to aid the user in the preparation and storage of a message. When a user is on vacation it is also possible that arrangements within the IPM UA could be made to allow mail to be forwarded to another nominated IPM UA. This type of extra functionality is outside the scope of the standardised X.400 message handling system as it does not encroach on the communication process. The auto-forward element specified in the IPM header is only to provide an indication that a message has been forwarded, rather than having an active part in the forwarding activity. It is in this way that X.400 systems can act as the messaging base for a system while the user interface can be built on top. This allows the user to interact with the type of facilities with which they are most familiar.

The IPM service has been defined, but what are its advantages over existing means of textual message-based communications?

Why Interpersonal Messaging?

In its general usage IP messaging will appear to be similar to existing private/public electronic mail systems with a certain amount of extra user-friendliness and flexibility in the means of addressing messages. Originator/Recipient (O/R) names will be based on similar information to that of the postal address. The kind of information found on a typical business card will generally be sufficient to uniquely specify the recipient within a country and within a certain organisation. Even less information than this may be required; it depends purely on the uniqueness of the address. This move away from machine oriented addresses should please users who have struggled to come to grips with unmanageable addressing schemes which some systems have employed in the past.

The actual content of an IP-message consists of one or more body parts and each of these can be of a different type. Examples of these might be text, facsimile or scanned graphics, voice, etc. (see Table 6.2). This kind of flexibility has only recently been available in electronic mail facilities offered in conjunction with integrated office systems, but it is not an option with public electronic mail systems.

Another aspect to an X.400 service is that the PTT operated ADMDs will offer conversion facilities (refer to Table 3.1) which will allow the sort of communication options indicated in Figure 6.4. Apart from communication to telematic services shown in this figure it has earlier been indicated that X.400 systems may act as a means of integrating and interconnecting other systems such as public and private electronic mail. The only proviso with message interchange between different services and systems is that the body

IA5 Text – character strings from the International Alphabet No 5 repertoire.

TLX – telex message content.

Voice – a bit string representing digitised voice.

G3Fax – Group 3 facsimile information encoding.

TIFO – A document encoded in accordance with the document interchange protocol for telematic services, recommendation T.73.

TTX – A body part consisting of a teletex document.

Videotex – A string encoded in accordance with the videotex encoding recommendations T.100 or T.101.

Nationally defined – Any type of body part defined by national agreement.

Encrypted – For text which is security sensitive and hence requires encryption.

SFD – Simple Formattable Document content type as defined in X.420.

TIF1 – A T.73 document conforming to TIF.1 application rules.

Table 6.2   Defined Body Types for IP Messages

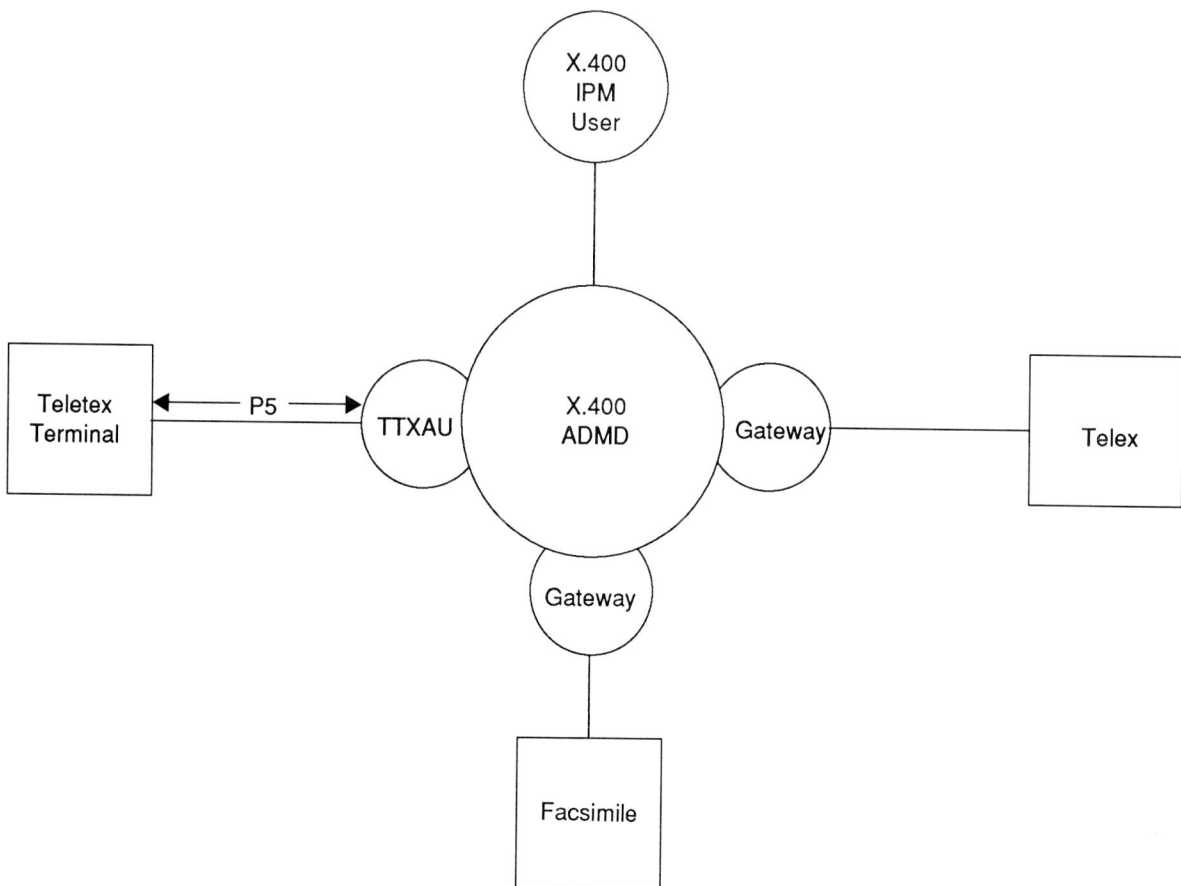

Figure 6.4   IPM Communication with Telematic Services

parts of the message have to be suitable for display on the receiving terminal equipment, eg a telex terminal cannot receive a scanned image.

Specific User Applications of Interpersonal Messaging

Although the general application of electronic mail/messaging has been mentioned so far, what will be the specific uses of the IPM service?

General mail

— messages for information only requiring no reply, eg a memo;

— messages requiring a quick reply, eg a sales enquiry;

— sales information for products and services.

Mail for specific communities

— for committee members, eg minutes from meetings, calling notices, committee papers, etc. Also potentially useful for fast voting on certain issues which do not require the committee to meet;

— closed user groups, circulars, notices of events;

— specific groups within an organisation, eg a certain department.

Interchange of processable documents (to be considered in detail in a later section):

— initially in the restricted format of the Simple Formattable Document (SFD) specification in CCITT X.420. In practice few will actually adopt this because of its restrictive nature;

— later a comprehensive service could be provided with ISO's Office Document Architecture, though this not permissible with the current recommendations;

— files structured in accordance with proprietory WP formats.

These are just some of the more obvious examples of specific applications of IPM which spring to mind, but as experience grows with X.400 IPM an infinite number of these will evolve.

Again it must be emphasised here that it does not matter what system or service the users are connected to; all can be accessed via an X.400 network utilising either gateways or access units. In this way the flexibility of X.400 systems should act as a catalyst which will heighten people's awareness of the potential of electronic messaging as an essential business communications tool.

PROCESSABLE DOCUMENT INTERCHANGE

In recent years the influx of word processors and PC-based word processors into the office environment has become ever more rapid. During this time, developments in equipment and the emergence of a far wider range of manufacturers has meant that the office has become littered with incompatible types of equipment.

After the initial steps in office automation – standalone devices – had been taken, it soon became apparent that the interconnection/integration of such systems was of paramount importance. It has already been indicated that in this context, the use of X.400 will be an immense step forward in terms of the basic integration/interconnection of existing systems and services. Linking office systems is only the first step in achieving a virtual system, that is a system which can present each individual user with an interface with which they are familiar. The user interface is that part of a system with which they will interact directly. One large stride towards a virtual system is the exchange of processable documents.

Processable document interchange implies that the features and structure which the originator has built into a document will be maintained when it is transferred through a system. A further requirement is that the recipient of the document should not only be able to view it, in the originator's intended format, but also should have the ability to alter or edit it as required.

Practical Aspects of Processable Document Interchange

For electronic text communications which are external to a specific organisation (ie inter-organisational) the transfer of text in its final form is generally all that is required. Depending on the type of document being transferred, ASCII text may be sufficient, or there can be a requirement for additional presentation featurers to be included. However, such documents will be in their final form and so there is no need to modify them when received.

For intra-business communications it has been shown that processable document transfer is a major requirement. Meeting this need will require substantial investment both monetary and in terms of development, but it looks likely that X.400 and some other standards may have an impact in this area.

The first stage of establishing processable document interchange is to provide the basic communications link, and in the context of this report X.400 is proposed because it has been internationally standardised and is achieving worldwide support. A further asset of an X.400 link is its ability to carry any form of structured or unstructured data, therefore offering great versatility. Once the basic communications link has been provided it is entirely feasible to transfer just the ASCII text of a document between dissimilar systems. However, the requirement is for processable document transfer, which will maintain all or most of the original's structural and presentational attributes, as defined by the author. The major problem is that dissimilar systems all generally use different methods for the creation and storage of documents and so the question of format conversion will arise. Format conversion can be simply described as mapping the features/facilities of one system to that of another.

Various solutions for document format conversion, on a one-to-one basis (eg DEC to WANG etc), have been and indeed are being used with some success. These include:

— black-box hardware conversion between systems, ie one format to another;

— software conversion routine—run on the user system to be invoked before document transfer;

— disk-to-disk conversion, either in-house or by a third party.

These offer a relatively simple solution to the problem provided that there is only a small population of different systems. But when there are a large number of dissimilar systems the number of conversions required escalates dramatically (see Example 6.4).

Example 6.4

When it is required to convert a document from one format to another, a conversion of some kind has to be carried out. In this example, the concept of a conversion unit (CU) is used which can represent any of the methods of document format conversion defined previously. Figure 6.5 illustrates this method for a population of three unique systems.

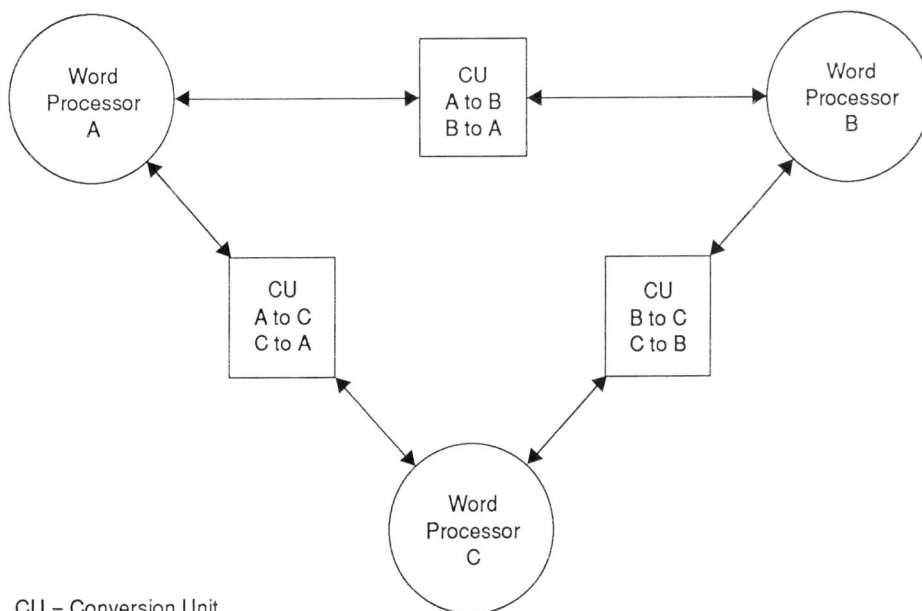

CU = Conversion Unit

Figure 6.5   The Use of Conversion Programs

This method of facilitating document transfer has two distinct problems:

(i)   For every extra unique systems linked the number of CUs will increase by N-1 (where N is the number of unique WPs). The cost of this approach grows exceptionally in relation to the number of unique systems.

(ii) If a design change or modification is made to one particular system in the network all its associated CUs (ie N-1) would also have to be altered.

It should now be obvious, from Example 6.4 that where the population of unique systems is significant that the straight format conversion solution will rapidly become both expensive and difficult to manage.

A fourth and more complex/expensive solution is to make all product purchases from a particular vendor who has proposed a specific architecture to which all their products should conform. Such a purchasing decision will have required both foresight and a substantial amount of co-ordination within a large organisation. The major problem with this route is that you are instantly locked into a single vendor domain for all future equipment purchases. Some time ago IBM saw the wisdom of this approach and proposed its Systems Network Architecture (SNA) within which document formats are standardised and a delivery system is provided. This has achieved some measure of success and IBM have published their standards in order to allow other vendors to produce conforming products.

The simplest solution to the problems defined so far is to propose an international standard for interchanging processable documents. This would act as an intermediary format to which everyone would convert their documents before carrying out transfer over the multi-purpose medium of an X.400 link. It would provide an unambiguous method of defining the features of documents and a clear format for the data stream to be interchanged. Reference to Figure 6.6 will indicate the effect of this concept on the interconnection of dissimilar systems.

It can be seen that each unique system has one CU associated with it. The CUs in this network only

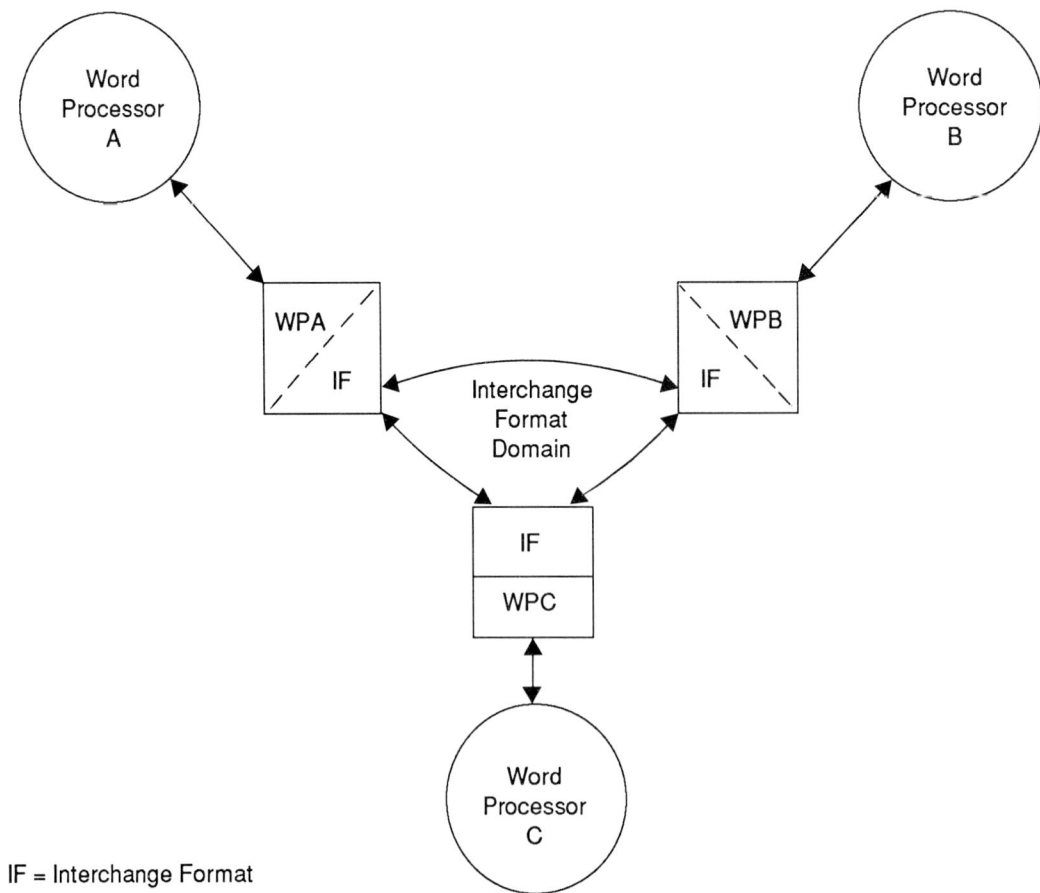

Figure 6.6   The Use of a Common Interchange Format

convert to and from the interchange format. Because only one CU is associated with each system domain the cost of enlarging the network or updating/changing a specific system is minimised.

Options for Processable Document Transfer

Within the original X.400 recommendations, X.420 contains a Simple Formattable Document (SFD) specification which offers a method of defining the attributes of a document in an unambiguous manner. SFD was devised to allow the interchange of documents which can be described as formattable but not yet formatted.

Differences between the physical characteristics of users' equipment, such as line length, vary quite widely. This means that information formatted in accordance with the originator's output device can appear almost unreadable when reproduced on the recipient's terminal. The problem can be solved by permitting the originator to specify only the logical structure of the document, ie the order of its specific objects such as title followed by paragraph 1 and then paragraph 2, etc. The detailed layout structure would then be left to the recipient's terminal to organise.

The SFD specification also allows attributes to be specified for each of the objects of a document, eg a heading which is to be in bold face. In this way the simple attributes, along with the logical structure of a document, are defined using SFD. A document coded in such a way can be thought as processable because the recipient's system will be able to interpret the features of it, as specified by the originator, and carry out modifications to the original format. As indicated previously, SFD body parts are permissible within IP-messages, so it is possible to carry out this simplified form of processable document exchange between IPM-UAs.

Improvements within today's text processing equipment has meant that documents can now have complex structures and presentation attributes. These are outside the scope of the SFD specification, which can only deal with very simple features. In an attempt to remedy this situation and to produce an all-purpose means of representing complex office documents ISO commenced work on their Office Document Architecture (ODA) standard. This work has now achieved International Standard (IS) status and looks certain to have a wide range of applications in tomorrow's office systems. In the existing 1984 X.400 recommendations ODA is not a defined body part; however, this will be remedied in the 88 version, which will take account of this requirement. It is possible to transfer documents formatted to ODA via systems based on the 84 X.400 recommendations by allowing them to assume a 'Nationally Defined' status as an IP-message body part. Another option is that the parties who wish to carry out ODA transfer formulate an agreement whereby anything specified as, say, SFD is accepted as ODA etc.

The Simple Formattable Document Specification

A Simple Formattable Document is comprised of character coded information only. Within such a document the text is organised into a logical sequence of paragraphs. The document and its individual paragraphs are referred to as specific logical objects. Each of the paragraphs is, in turn, a series of content portions, only one of which character-text is defined at present. The basic SFD structure is indicated in Figure 6.7.

The following are more formal definitions of the components of an SFD:

Specific logical objects:

— document – A document is a logical sequence of zero or more paragraphs;

— paragraph – A paragraph is a logical sequence of zero or more content portions.

Content portion

These are the portions of a document which are to be output during the rendering process. They convey what is to be rendered rather than how. So far, only one type of content portion has been defined— character text.

Character text

A character text content portion is a logical set of zero or more graphic characters as defined in CCITT recommendation T.61 (see Figure 6.8).

```
                              ┌───────────────┐
                              │   Document    │
                              └───────────────┘
                    ╱                 │                 ╲
                   ▼                  ▼                  ▼
          ┌───────────────┐  ┌───────────────┐  ┌───────────────┐
          │  Paragraph 1  │  │  Paragraph 2  │  │  Paragraph 3  │
          └───────────────┘  └───────────────┘  └───────────────┘
                  │                  │                  │
                  ▼                  ▼                  ▼
          ┌───────────────┐  ┌───────────────┐  ┌───────────────┐
          │   Character   │  │   Character   │  │   Character   │
          │     Text      │  │     Text      │  │     Text      │
          └───────────────┘  └───────────────┘  └───────────────┘
```

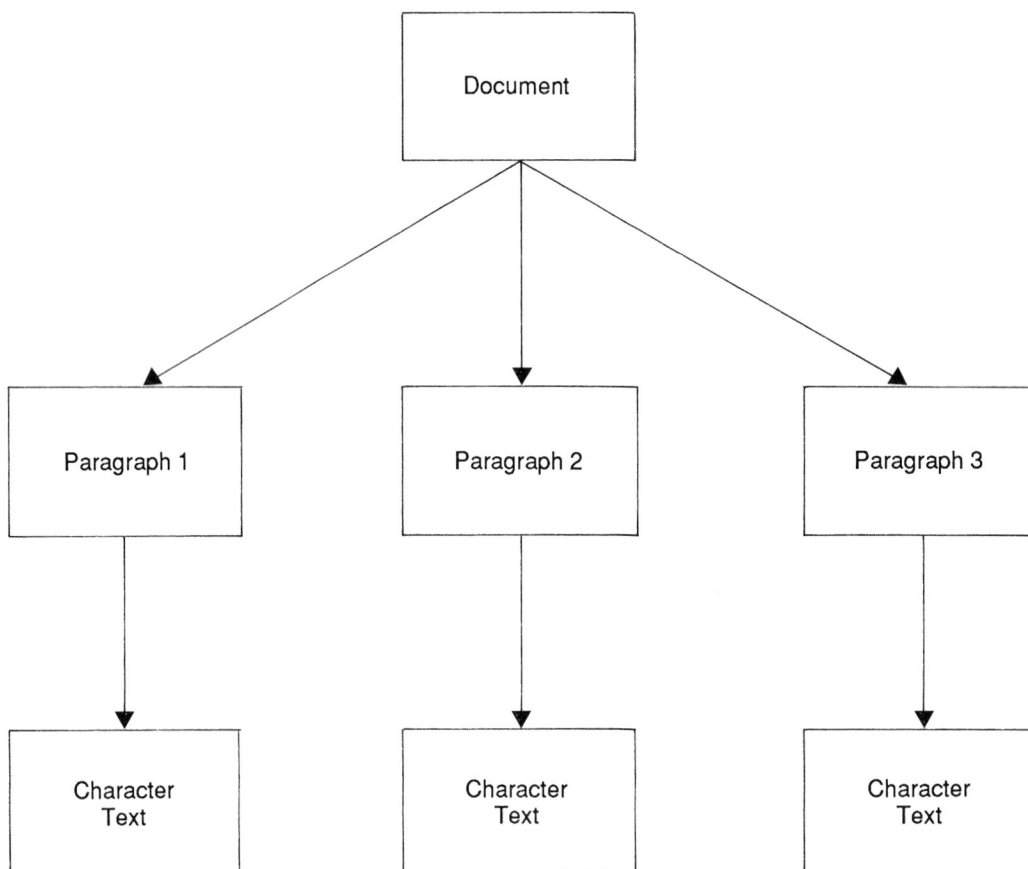

Figure 6.7   The Basic Structure of an SFD

The specific logical objects of a document have one or more attributes, which fall into one of two categories, either variable or unspecified.

Variable attribute

These can be specified by the originator of the document and have to be maintained during the rendering process.

Unspecified attribute

These are under the control of the rendering entity and not the originator.

Figure 6.9 shows examples of the attributes which are attached to the various protocol elements.

Simple Formattable Document—Document Interchange Format (SFD—DIF)

When a SFD is transferred between systems it will be represented as a sequence of octets or bytes. The structure of this sequence is called the SFD-Document Interchange Format (SFD – DIF) and it is this which governs the SFD body part of an IP message. SFD – DIF conforms with the presentation transfer syntax rules as defined in recommendation X.409, which provides a standard notation for describing complex data structure.

Although the detailed structure of SFD – DIF is outside the scope of this book it is of interest to look at what is termed the structured encoding of a SFD. This can be produced at the originator's terminal in order to indicate exactly what featurers of the document will be transmitted to the recipient. The type of features considered are the paragraph boundaries and the attributes which are attached to the various specific logical objects. A sample document and its structured encoding is shown in Example 6.5.

b8	0	0	0	0	0	0	0	0	1	1	1	1	1	1	1	1
b7	0	0	0	0	1	1	1	1	0	0	0	0	1	1	1	1
b6	0	0	1	1	0	0	1	1	0	0	1	1	0	0	1	1
b5	0	1	0	1	0	1	0	1	0	1	0	1	0	1	0	1

b4 b3 b2 b1	row	0	1	2	3	4	5	6	7	8	9	10	11	12	13	14	15
0 0 0 0	0			SP	0	@	P		p				°			Ω	ĸ
0 0 0 1	1			!	1	A	Q	a	q			¡	±	`		Æ	æ
0 0 1 0	2			"	2	B	R	b	r			¢	²	´		Đ	đ
0 0 1 1	3			④	3	C	S	c	s			£	³	^		ª	ð
0 1 0 0	4			④	4	D	T	d	t			$	×	~		Ħ	ħ
0 1 0 1	5			%	5	E	U	e	u			¥	µ	¯			ı
0 1 1 0	6			&	6	F	V	f	v			#	¶	˘		IJ	ij
0 1 1 1	7			'	7	G	W	g	w			§	·	˙		Ŀ	ŀ
1 0 0 0	8			(	8	H	X	h	x			¤	÷	¨		Ł	ł
1 0 0 1	9			)	9	I	Y	i	y				②			Ø	ø
1 0 1 0	10			*	:	J	Z	j	z				°			Œ	œ
1 0 1 1	11			+	;	K	[	k				«	»	¸		º	ß
1 1 0 0	12			,	<	L	\	l	|				¼	③		Þ	þ
1 1 0 1	13			–	=	M	]	m					½	˝		Ŧ	ŧ
1 1 1 0	14			.	>	N		n					¾	˛		Ŋ	ŋ
1 1 1 1	15			/	?	O	①	o					¿	ˇ		'n	

Figure 6.8   Code Table for Graphic Characters Showing Space in Position 2/0, the Primary Set of Graphic Characters in Positions 2/1 to 7/14, and the Supplementary Set of Graphic Characters in Positions 10/1 to 15/14.

Source: CCITT REC. T.61

Protocol Element	Variable Attributes	Examples of Unspecified Attributes
Document (a logical object)	Page heading	Output medium dimensions Top, bottom, left and right margins Presence and position of page numbers Number of columns per page How page breaks are chosen
Paragraph (a logical object)	Alignment Left indentation Bottom blank lines Graphic rendition	Line height How line breaks are chosen
Character text (a content portion)	Graphic rendition	Font Pitch

Figure 6.9   SFD Protocol Elements and their Attributes
Source: CCITT REC. X.420

Example 6.5

Basic Document:

The establishment in various countries of telematic services and computer-based store-and-forward message services in association with the public data networks creates a need to produce standards to facilitate international message exchange between subscribers to such services.

The CCITT, considering:

(a)   the need for interpersonal messaging and message transfer services;

(b)   the need to transfer messages of different types having a large variety of formats;

The Structured encoding of This Basic Document:

```
specificLogicalDescriptor {document, {
    defaultValueLists {paragraphAttributes {alignment justified}}}},
specificLogicalDeescriptor {paragraph, {}},
textUnit {{},
    "The esablishment in various countries of telematic services and computer-based store-and
    forward message service in association with public data networks creates a need to produce
    standards to facilitate international message exchange between subscribers to such services.},
specificLogicalDescriptor {paragraph, {}},
textUnit {{},
    "The CCITT, considering:"},
specificLogicalDescriptor {paragraph, {
    presentationDirectives {alignment leftAligned},
    layoutDirectives {leftIndentation 5}}},
textUnit {{},
    "(a) the need for interpersonal messaging and message transfer services;"},
specificLogicalDescriptor {paragraph," {
    presentationDirectives {alignment leftAligned},
    layoutDirectives {leftIndentation %}}},
textUnit {{},
    "(b) the need to transfer messages of different types having a large variety of formats;"}}
```

ISO OFFICE DOCUMENT ARCHITECTURE (ODA) STANDARD

Work initiated in 1977 on Open Systems Interconnection (OSI) by the International Standards Organisation (ISO) is now coming to fruition with the implementation of application-oriented communication services. One of these is a new International Standard for Text Interchange, the 'Office Document

Architecture and Interchange Format' ISO IS8613. This Office Document Architecture (ODA) Standard, as it is more often called, facilitates the interchange of many types of documents between dissimilar document processing systems. It does this by providing an independent model for representing the contents and logical layout structure found in typical office documents. It also provides an Interchange Format which specifies the way in which the document is to be encoded as a data stream.

As mentioned above, ODA now has International Standard (IS) status and so can be considered technically stable and ripe for implementation. Vendors are now considering how to incorporate ODA into their existing and future document processing systems.

The purpose of the standard is to facilitate the interchange of office documents, providing for their representation in such a way as to enable the documents to be reproduced and processed by the recipient, where interchange is by means of data communications or the exchange of storage media, eg floppy disks. The ODA Standard caters for a number of office documents such as memoranda, letters, forms and reports. Documents may include both text and pictorial information by allowing character box, geometric and photographic elements all within a single document.

Office Document Architecture

The model detailed within the ODA Standard describes the document in terms of a 'document profile' and 'document body'. The profile is used for document management, and it specifies characteristics which apply to the document as a whole. Such characteristics would be filing and retrieval details, security aspects, external references, the type of document content and its structure. The reason for the use of a document profile is that it can be sent ahead of the document body for the receiving equipment to ascertain whether or not it wants or is able to receive the document body. This obviously prevents the needless overhead of complete document transfer when it is not required or not possible.

The ODA Standard provides the definition of an abstract model for representing an office document body and layout feature. The model is organised into two hierarchial structures, the logical structure and the layout structure, which are linked by layout directives.

The content of the document consists of character box, geometric or graphic elements. The logical structure relates these to logical text objects such as headings, sections, paragraphs, figures and footnotes. The layout structure relates the content to its positioning and rendition on the presentation media; the layout directives express the layout requirements of the logical objects.

Documents can be regarded as being members of different classes, examples being memorandum, letter or report, where a particular class can be defined by specifying a common set of properties. These properties can be types of logical text objects that occur in a specific manner throughout the document; for example, header and footer text that appear on every page or the way in which a chapter begins, eg chapter number, then the title etc. By having such a logical structure for a document consistency of structure is maintained within a specific class.

Those structures which are common to a specific class are called generic logical and layout structures, the relationships between them being generic layout directives. Structures which relate to a particular instance within a class are termed specific logical and layout structures, the relationship between them being specific layout directives. In the following sections each of the four possible structures are considered separately.

Specific Logical Structure

Logical structures describe how a document may be decomposed into chapters, sections, subsections and paragraphs. The specific logical structure is a tree structure of logical objects. Relationships between the logical objects as well as other additional characteristics are specified as attributes of the objects. An example of the structure is shown in Figure 6.10.

Generic Logical Structure

Generic logical definitions describe the general hierarchical structure for all documents within a particular document class, the possible sequences of the branches of the permitted tree structure; the type of logical object within the structure which are known as object classes; and the common content portion. An example of this structure is shown in Figure 6.11.

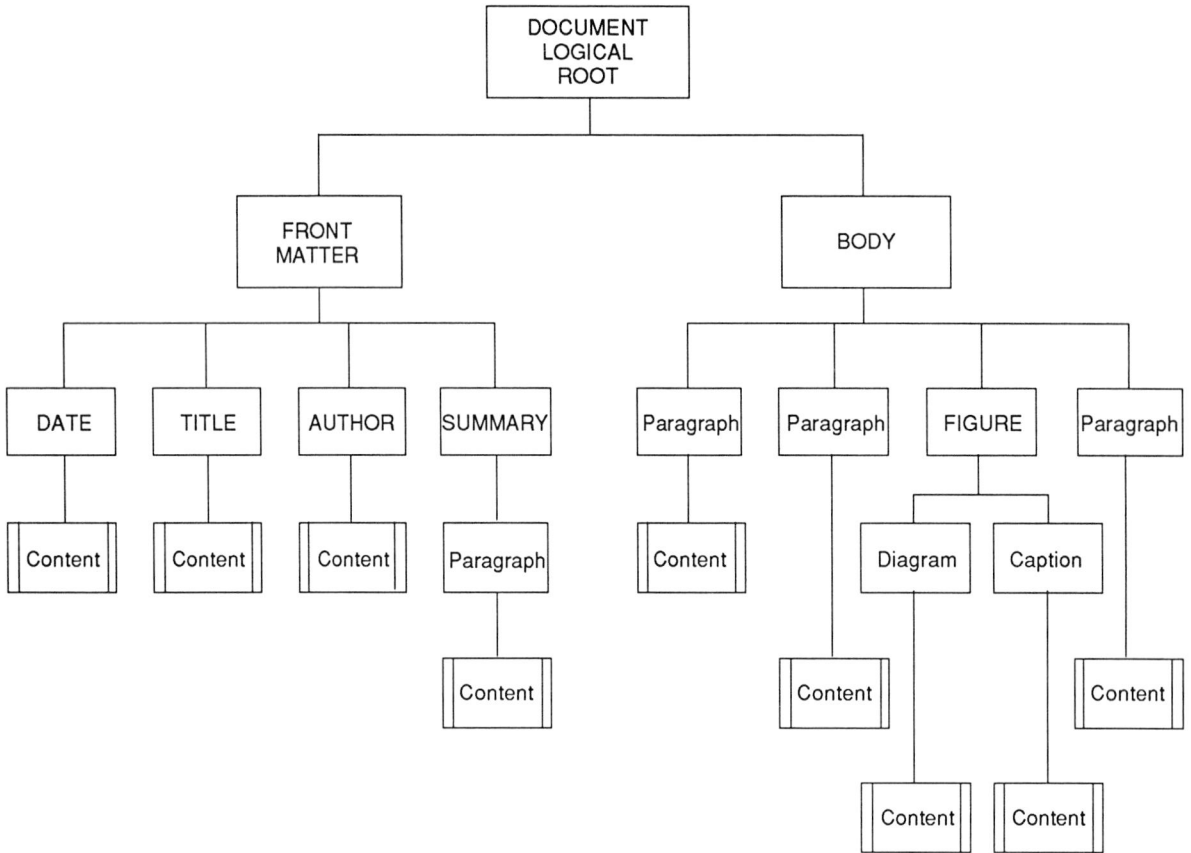

Figure 6.10   Example of a Specific Logical Structure for a Simple Report

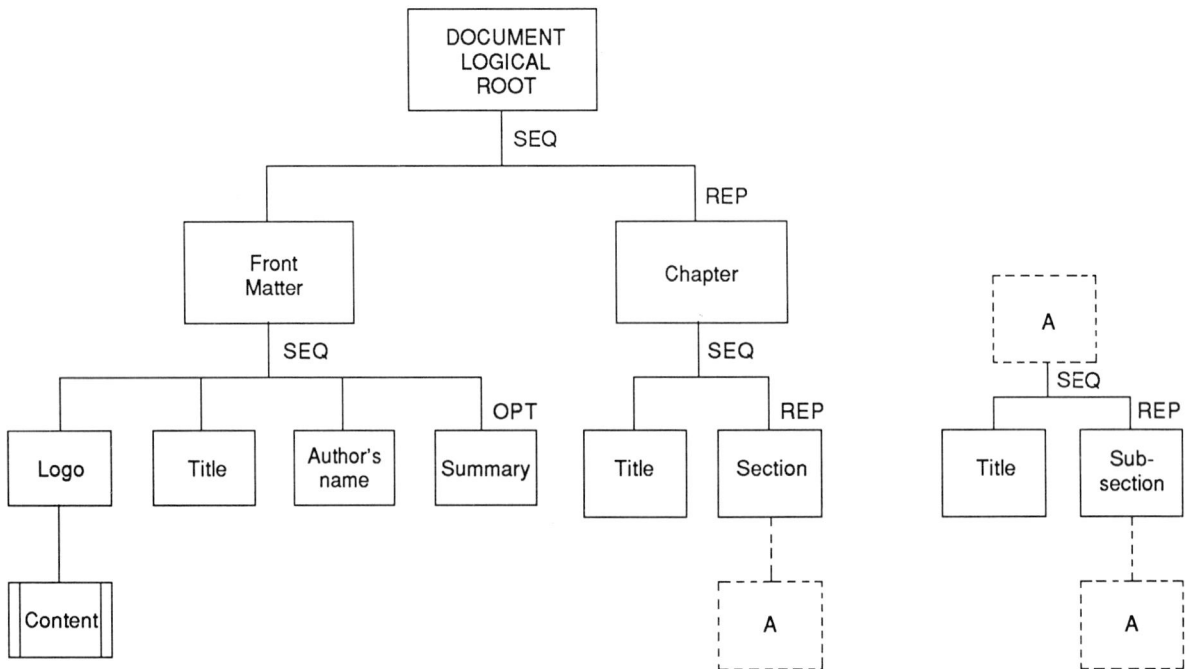

Figure 6.11   Example of a Generic Logical Structure

Specific Layout Structure

The specific layout structure mainly describes the positioning of the content portions and their visual attributes on the rendition surface. It will be specified as a tree structure of layout objects. The possible layout objects are called page, sets, pages, frames and blocks. These provide different kinds of representation for the parts of the document image. Page sets describe a set of correlated pages with similar layout characteristics. A block is directly related to a content portion of a document.

Generic Layout Structure

Generic layout definitions describe the rules for establishing the layout structures of all documents within a particular document class. They also detail any attributes of the layout objects and common content portions from a document class. An example of this structure is shown in Figure 6.12.

Office Document Interchange Format (ODIF)

This part of the standard describes the mechanisms for exchanging documents which are structured in accordance with the architecture. It details the way in which the structured document is to be formatted in a data stream.

The interchange of a document is performed by first transferring the document profile and then the document body. For interchange purposes the document body is structured into so-called descriptions (each specifying a generic/specific logical/layout object) and text units (each specifying a content portion).

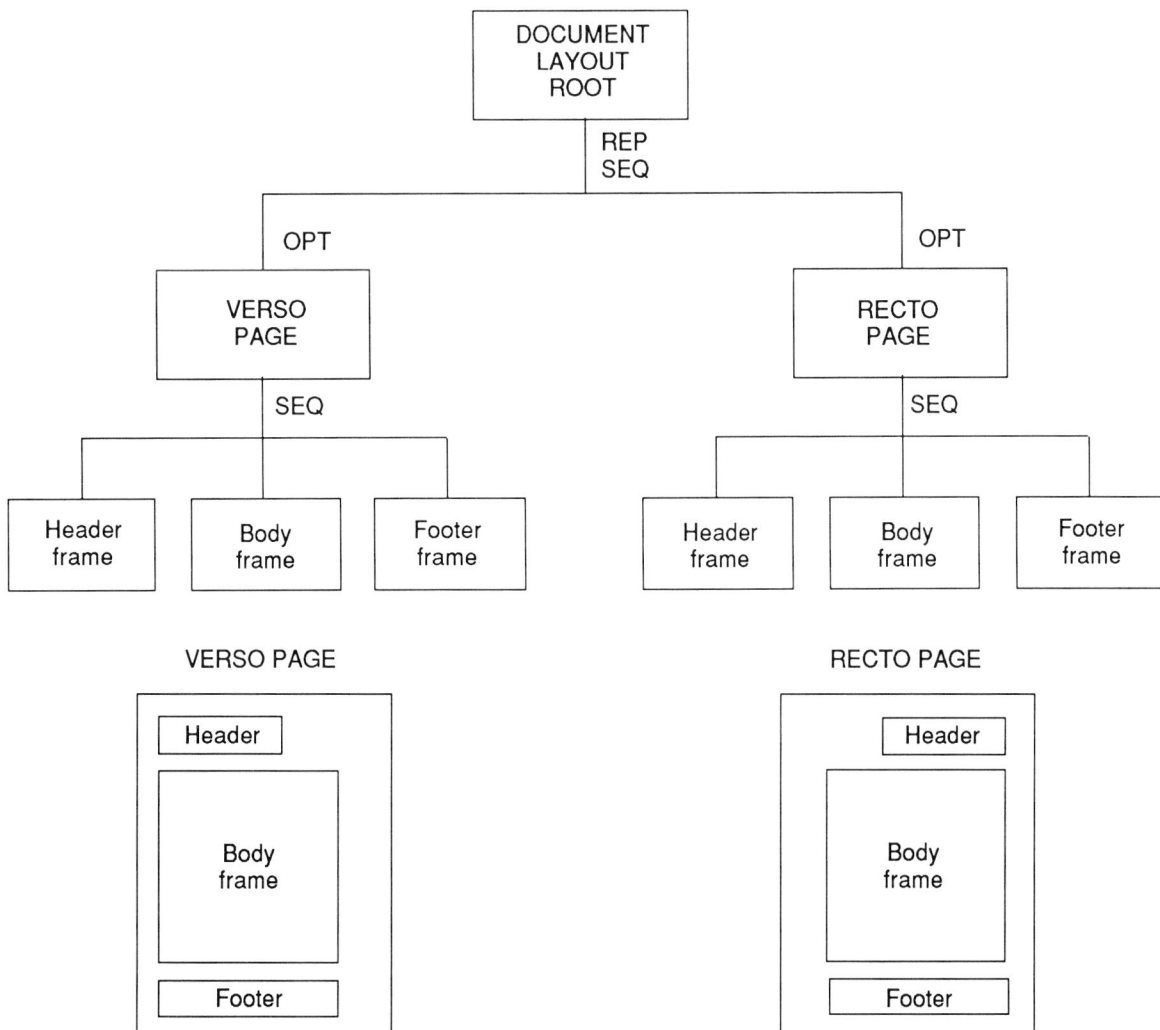

Figure 6.12 An Example of a Generic Layout Structure

The descriptions and text units may be interchanged in two different orders, specifying two distinct possible data streams.

Data Stream A

1 Document profile

2 Description of generic layout definitions

3 Description of generic logical definitions

4 Text units for common content portion

5 Description of specific layout objects

6 Description of specific logical objects

7 Specific text units.

This type of data stream can always be used as it contains all the required information about the document.

Data Stream B

1 Document profile

2 Generic 'tree'

3 Specific 'tree'.

This data stream may be used if only the layout part or only the logical part of a document specification is to be exchanged.

From the discussion so far it should be obvious that the application of processable document interchange via X.400 systems has certainly not been neglected. There are two standardised methods for carrying out such a transfer using either the Simple Formattable Document (SFD) specification or the forthcoming Office Document Architecture (ODA) standard from ISO. The former will only really be of use in the most elementary of cases while the latter, ODA, has been designed to cope with all the possible options in tomorrow's composite documents, such as text, voice, graphics.

Another option is to carry out some form of non-standardised processable document interchange and, because of the flexibility of X.400 systems for all-purpose data transfer, this is entirely feasible. This situation can prevail for intra-company communications providing that all groups within the company agree on a specific format. There are, however, a number of problems that have to be considered before traversing this particular route:

— There is the problem of which particular solution to choose.

— Will this solution tie the company to products from one manufacturer, thereby defeating one of the major advantages of OSI systems?

— In the event of a merger situation between established companies, with established networks, how will these be able to integrate their individual processable document transfer applications?

These problems can be overcome by conforming to an internationally standardised solution such as ODA. It is also likely that ODA is probably more far-sighted in its approach to these problems and so will not become obsolete in a short period due to a lack of flexibility.

ELECTRONIC DATA INTERCHANGE OVER X.400

The Need for Electronic Business Data Interchange

In the past, production of documents for trade was purely manual. A person would either fill in a form by hand or produce it on a typewriter. But in recent years many businesses have automated this process through the use of computers, producing large numbers of documents from data held in their memory banks. This automation, however, generally falls short of completing the full electronic cycle of production and transfer of business data.

Most businesses are involved in trading their goods and services with other companies or individuals and this involves transferring information via the outdated postal service. Even when information is held on computer systems, ranging from mainframe to personal computers, it still has to be produced in hard copy format in order that it may be transferred, and at the receiving end the information will undoubtedly have to be rekeyed into the recipient's system. This is due to the incompatibility of the computing systems that prevail in today's office environment. Obviously this paper generation and rekeying of data causes substantial overheads in organisations and means that the process is also subject to human error.

Recent figures indicate that 7.5 – 15 per cent of international trade invoice values are lost worldwide because of errors in documents and inefficient data transmission – equivalent to lost revenue in the order of $40 billion a year. Research in the UK has shown that 50 per cent of all letters of credit are rejected by British banks at their first presentation because of errors or because they are late. In some cases errors which are not readily identifiable can cause severe problems in over/under ordering and late delivery of goods and services. Such situations are difficult to satisfactorily when they reach this stage. Further research has shown that the costs involved in the manual production of a letter, ie secretaries, internal postal staff, etc, can often be up to fifty times the value of the postage stamp.

Any means of ensuring the reliable transfer of such data and reducing the associated overheads will be beneficial to any business and aid their ability to trade in the international marketplace. The obvious solution to this problem is to electronically transfer information directly between the computer systems of trading partners. In this way excessive human intervention, paper generation and postage problems/cost will be avoided. But it has already been stated that most computing systems are not able to be directly interconnected, so some form of standardised method of interconnection is required.

Potential Benefits of Business Data Interchange

In 1984 Istel Limited and the Institute of Physical Distribution Management (IPDM) conducted a joint survey looking into the potential for inter-business data communications. The results of this report give some insight into both the volume of paper-based communications between organisations and the need, which most of them felt, for the improved information flows resulting from the electronic data communications.

Reference to Figures 6.13 (a) and 6.13 (b) shows that in all cases the industries surveyed utilised the postal service as their main form of document transfer. In fact, 60 per cent were being transmitted in this manner, whereas the utilisation of electronic interchange represented only a small portion of the total.

The survey also assessed the average costs of document flows in 1984. These were:

	Purchase orders orders	Sales invoices	Supplier invoices	Sales Invoices
Average cost	£12.46	£8.52	£5.86	£3.87

Even allowing for the fact that these figures are from 1984 it is easy to see the potential for cost saving, and the situation will be even more apparent now. When questioned about whether or not they felt that their business communications would improve through the use of EDI the results in Figure 6.14 were obtained. These show conclusively that most organisations felt that they would benefit from the improved information flows achieved by the use of EDI.

Subsequent experience with Istel has shown that inter-business data communications can reduce the direct costs of document processing by up to 25 per cent and it can also aid in the reductions of indirect costs.

(All figures reproduced from Istel introduction to EDICT and the joint IPDM-Istel report on data communications.)

Introducing the Concept of Electronic Data Interchange

Most companies that have to process large or small amounts of data now do so via computers, and these help greatly in the storage, retrieval and day-to-day manipulation of information. However, lack of compatibility between such devices means that when data is to be transferred from one organisation's computer to that of another the outdated postal or courier services must be relied on. This has

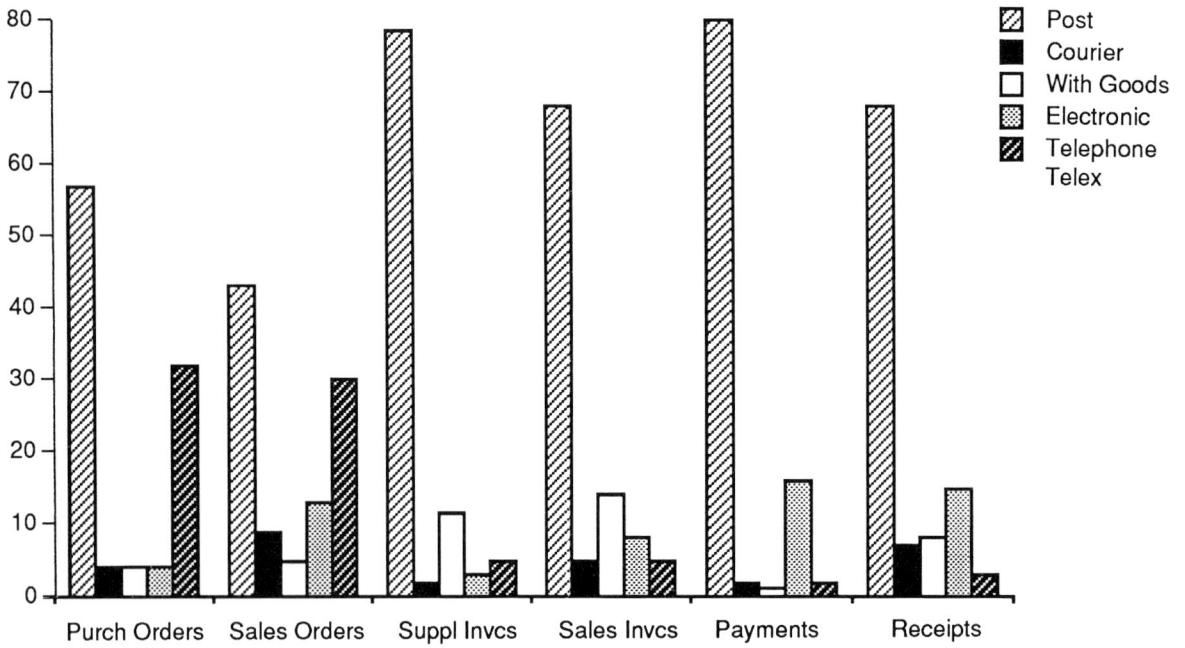

Figure 6.13(a)   Methods of Transmitting Documents

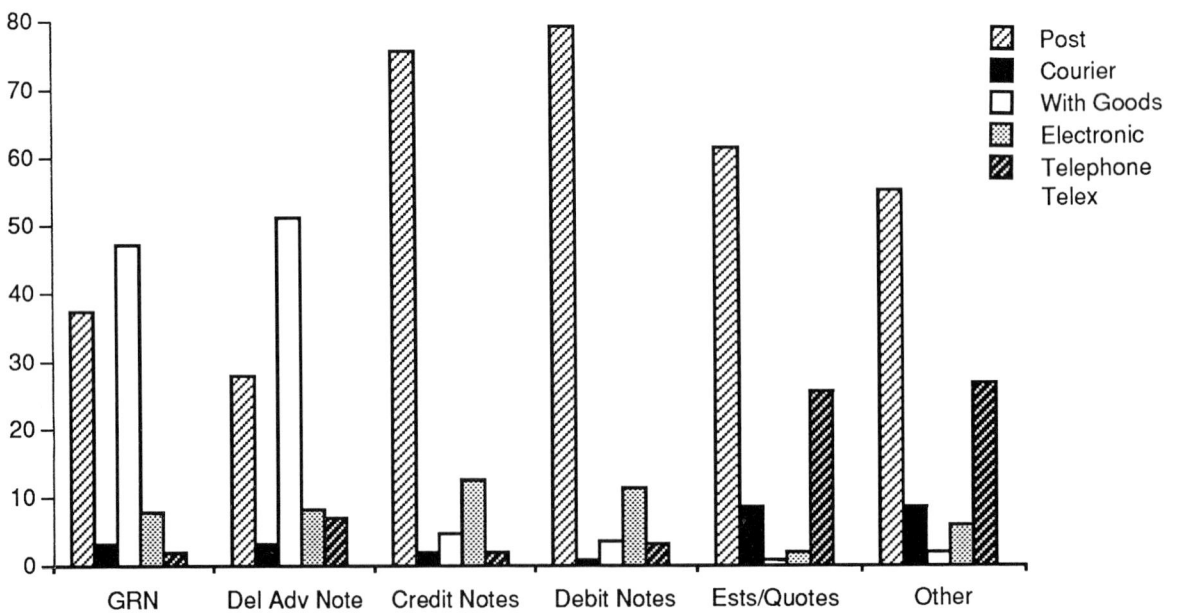

Figure 6.13(b)   Methods of Transmitting Documents

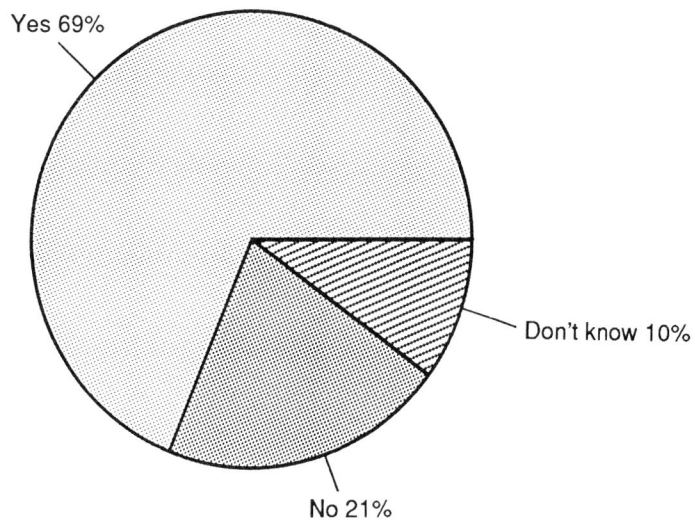

Figure 6.14   Would Organisations Benefit from Using EDI?

disadvantages in two areas; the time factor involved in such a transfer process – which can be lengthy – and the introduction of human error to the process when data is rekeyed on the recipient's system. This cycle is indicated below in Figure 6.15.

The obvious solution to these problems is to establish an electronic link between the computer systems of trading partners, thus overcoming the time delay factors and eliminating human intervention from the transfer – always a potential source of errors. While this is acceptable in theory it proves to be difficult in practice unless the computer systems have common data and file formats or employ some form of conversion, either 'black box' or bureau service. Even when this is the case there is still a decision to be made as to which specific means of information transfer will be used, ie which file transfer protocol and link. The best way is to have compatibility between data and file formats of systems, and this is the general aim of Electronic Data Interchange—or EDI—to use the common acronym.

The generally excepted definition of EDI is:

'the interchange of structured data, by agreed message standards, from computer to computer, by electronic means'.

The key words here are 'agreed message standards'. These will specify the manner of expression, the type of data elements to be transferred and the means of structuring these elements into messages, groups

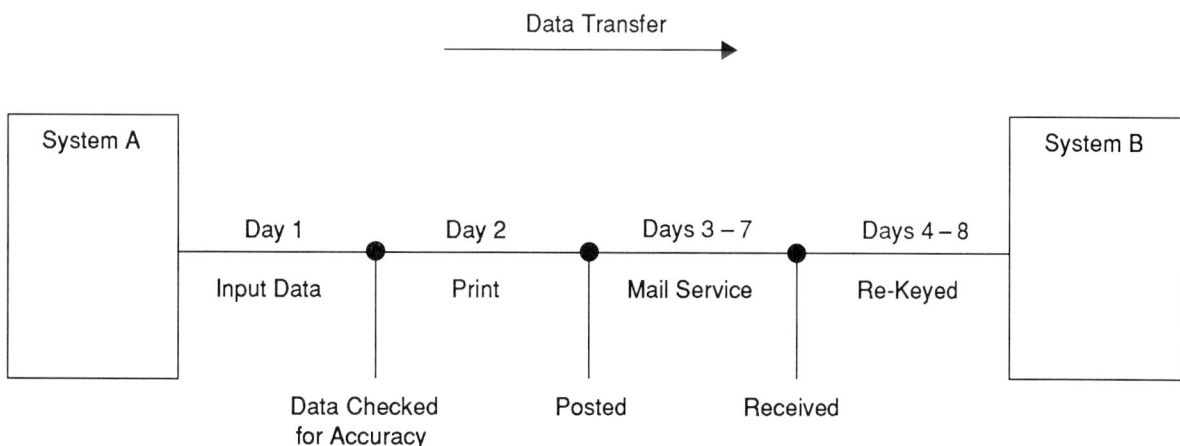

Figure 6.15   A Typical Cycle in the Postal Transfer of Data Between Trading Partners

of which will form files or batches of information to be interchanged. The messages themselves will be formed with a specific task in mind and would generally have a relation to paper-based documentation, such as invoices, purchase orders, credit notes, etc, which they will eventually replace.

It should be mentioned that the above definition does exclude the EDI VADS services which have emerged. These services allow users to send data in a standard EDI format to the service which then places the interchange in the intended recipients mailbox. This approach effectively de-couples the trading partner from the communication process.

All EDI standards which have evolved so far have ensured that they achieve complete independence from the method of transport being used. This means that a file formatted in accordance with an EDI standard can be equally well transferred electronically via a file transfer protocol or as information stored on a floppy disk or magnetic tape. Even when the latter means of physical media transfer is employed, EDI standards still overcome the introduction of additional human error to the process, although the time delay factors might still apply. Utilising the concepts of EDI we can look at its effect on the delays involved in data transfer between systems.

Figure 6.16 indicates the connection of two computer systems directly. In practice, however, this proves to be difficult to achieve. Not only do individual computer systems use their own data formats, they also use their own communication protocols. Because historically organisations have purchased such equipment from a wide range of sources, these communication protocols are also generally incompatible. The emerging Open Systems Interconnection (OSI - see Appendix 4) communication framework and protocol standards are meant to overcome such problems. In this context X.400 and FTAM, being OSI based, would then seem to be suitable applications to achieve the transfer of EDI information. FTAM's suitability for EDI applications becomes more apparent when considering bulk or batch transfer, whereas X.400 would seem to be most suited to smaller file lengths which need to be transferred to a large number of destinations. It is this latter area of EDI which should see rapid growth in the future as its use becomes more commonplace, that is, transfer of data not only with regular business partners.

The Benefits of Electronic Data Interchange

It is important at this point to attempt to identify the benefits of utilising electronic data interchange (EDI) as the basis of business information communications. It is only via this approach that an insight can be gained into the importance of this potential application for X.400 systems.

The major benefits of EDI on an 'industry wide' basis can be summarised as follows:

— reduction in the cycle time of documentation and the orders themselves;

— improved customer response and hence customer image;

— reductions in the direct and indirect costs of processing documentation;

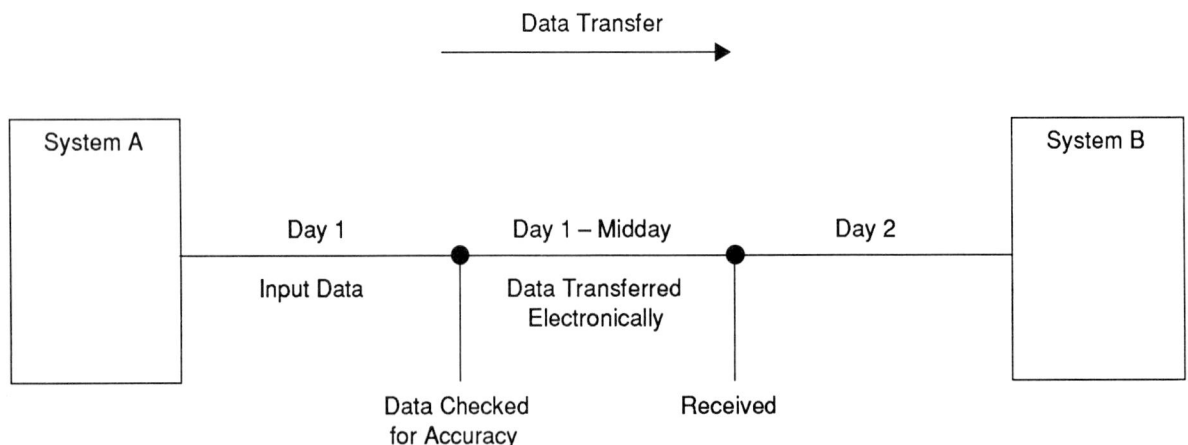

Figure 6.16   The Data Transfer Cycle Employing the Concepts of EDI

— lower incidence of errors in data;

— improved communication with suppliers;

— improved security.

Introduction of the Standardisation of EDI

The standards associated with EDI are independent of the medium used for transmission, and in OSI context refer to an activity above the application layer, often termed information standards. This allows data structured in accordance with any agreed syntex to be carried out utilising any available OSI application, eg X.400 or FTAM.

There have been numerous attempts to define sets of rules or standards to govern the electronic transfer of trading information, but the aspects which have to be considered have remained the same:

— *A set of defined data elements.* The items of information to be conveyed between trading partners, eg product codes, expiry dates, address of consignor and consignee, etc.

— *Syntax rules.* These describe the way in which the elements of data are formed into segments and batches before transmission.

— *Message design rules.* To allow the elements and segments of data to be grouped together in order to form standard types of message. These would ensure that messages are produced in a consistent and coherent manner.

— *Standard messages based on the design rules.* Standard types of messages for specific business transactions, eg invoice, purchase order, delivery instructions, etc. These would generally have a close relation, in terms of the information contained, to the paper documents they replace.

We now look at the bodies involved in the standardisation of EDI and the work they have completed to date.

The Groups Involved in the Standardisation of EDI

The Simplification of International Trade Procedures Board (SITPRO) EDI is not a new concept. As early as the 1970s the Simplification of Trade Procedures Board (SITPRO), a UK government sponsored body, produced its guidelines. SITPRO's work was then used as the basis for the European standards, UNTDI (United Nations Trade Data Interchange) and TDED (Trade Data Element Directory—since adopted by the ISO as IS 7372), produced by the UN Economic Commission for Europe (UN-ECE). In the United States interest in EDI has focused on the American National Standards Institute (ANSI) and their X12 work, which consists of:

— transaction set standards (messages);

— a data element dictionary;

— transmission control standard (syntax).

ANSI X12 has been widely adopted throughout America and North America but has failed to penetrate outside that continent.

In an attempt to produce a global standard for EDI the UN-JEDI group (UN joint EDI group) was formed. Its members were drawn from the following countries and organisations:

Belgium	Sweden
Canada	United Kingdom
F R Germany	United States
France	CEC
Netherlands	IATA
Poland	

The first task of the group was to produce appropriate syntax rules which were to be a combination of the work of the UN and ANSI, pulling together the areas of common ground and making compromises where none existed. Their efforts have been extremely successful and the syntax rules have already been accepted by ISO as an International Standard (via the Fast-Track procedures). This project has taken about two years from start to finish and indicates some of the enthusiasm and support for this topic.

Summarising the work of the UN-JEDI group so far:

— *The Trade Data Element Directory (TDED)*. A straight adoption of the ISO 7272, although it is likely that this will eventually be redefined to encompass the concepts of qualifiers, eg DATE

> (1) delivery
>
> (2) expiry
>
> (3) despatch etc.

Date is a generic type which is qualified for specific applications by different codes.

— *The Syntax Rules*. Developed from the UNTDI European standard and the ANSI X.12 American work. These syntax rules are now complete and issued as an ISO standard, IS 9735 called EDIFACT (Electronic Data Interchange for Administration, Commerce and Transport).

— EDIFACT Message Design Guidelines. These allow groups engaged in designing new message or modifying existing messages, to do so in a consistent manner which will enable other users to understand them.

— EDIFACT Standard Messages. The first one to be defined is the Specification for UN—ECE Standard Electronic Commercial Invoice Message.

In its position as the world standard for EDI, EDIFACT is the only real alternative when looking towards international trading. Because of this prominent position EDIFACT is worthy of some further consideration in order to give an introduction to the structure of its sytax. This exercise will also prove to be a useful introduction to the topic of EDI syntaxes in general, as they employ similar concepts.

EDIFACT Syntax Tutorial

ISO 9735 EDIFACT is quite a short document which specifies, in a condensed form, the rules for structuring user data and its associated service information to facilitate exchange. The approach taken in this tutorial will be to look briefly at the elements of an EDI transmission and then to see how these elements are formed together for the purpose of interchange.

As with any standard, EDIFACT is riddled with its own jargon which needs definition and clarification. The following are the major elements within an EDIFACT interchange:

— Data Element is an atomic unit of information which in general corresponds to the type of items of information found within trade documentation: possible examples would be invoice number, date of production, postal code, country code, bank branch name, etc. These are at present defined along with their number codes and mnemonics in the Trade Data Element Directory (TDED) ISO 7372. Data elements are analogous to a vocabulary, but for EDI rather than speech.

— *Composite Data Element*. A data element containing two or component data elements, eg communications contacts communications number and communications number qualifier in this case three components. Each item within a composite data element is separated by a ' : ' character which is a component data element separator.

— *Data Segment*. A data segment is the intermediate unit of information in a message. It will consist of predefined sets of functionally related data and composite data elements. An example might be a name and address segment which might consist of the following simple elements:

> Place/Party Function Code
>
> Place/Party Identification
>
> Name and Address
>
> Street Number

City Name

Postal Code

etc

Within a segment the various elements will be separated by a '+' character;

— Message. An EDI message is a collection of data, logically grouped into segments that will be exchanged to convey information, relating to a specific transaction, task or subject, between partners engaged in EDI. Messages will generally have some relation to the paper-based documentation or the trade functions which they are replacing. Elements of such messages might bear a very close relationship to those of a paper invoice or credit, for example. Each message will always begin with a special message header segment (UNH) and end with a message trailer segment (UNT). The type of transaction to which the message is referring will always be specified in the UNH segment.

— Functional Groups. These are groups of messages of the same type, eg purchase orders to one company but referring to individual transactions. Functional groups also have special header and trailer segments, UNG = Functional Group Header and UNE = Functional Group Trailer;

— Other Service Segments

UNA = Service String Advise, enables the definition of delimiters and indicators for a specific interchange;

UNB = Interchange Header, provided so as to start, identify and specify the interchange;

UNZ = Interchange Trailer, provided to end and check the completeness of the interchange.

Thus armed with a knowledge of the elements involved we can now look at the overall makeup of an EDIFACT transmission. Figure 6.17 is an extract of the actual EDIFACT standard.

In order to conclude this tutorial on the EDIFACT syntax and to relate it to the more familiar paper-based trading methods refer to Figure 6.18. This shows an example of a paper-based document in which elements and sections are boxed and identified in EDIFACT terminology.

IMPLEMENTING EDI – THE TECHNICAL OPTIONS

When implementing EDI the decisions to be made fall neatly into three basic categories: administrative, technical and operational. In trying to establish the relationship between X.400 and EDI, it is useful to assess the technical options for implementing EDI so as to gain an insight into the way in which EDI is being implemented now. In this way it will be easier to contrast the X.400 approach in subsequent sections.

The technical decisions to be made when implementing EDI fall broadly into three categories, these are:

— standards for data format;

— software interface;

— communications.

The first two considerations generally result in the adoption of standards for the EDI data, structure and messages, and a software interface to implement these. This will allow a common data interchange format between the trading partners. It has already been indicated that such an approach relieves the trading partners of any dependency on specific machine data formats. The alternative to this is to use conversion routines to convert incoming data structures to one which is compatible with the in-house system. The problem with this approach is that each trading partner has to maintain a separate conversion process for every type of system with which they deal. It is this very problem which has prompted the development of EDI standards to be used as a kind of generic interchange format between systems.

Once an organisation has decided to use an EDI standard the remaining problem is to choose one. This decision will be made including consideration of aspects such as: who you do business with, the applicability of message designs, the long term future, and potential migration of the current standard versions. When a decision has been taken on which EDI standard to use, the next step is to address the

A CONNECTION contains one or more interchanges. The technical protocols for establishment, maintenance and termination, etc, are not part of this standard.

An INTERCHANGE contains:
- UNA, Service String Advise, if used
- UNB, Interchange Header
- Either only Functional Groups, if used, or only Messages
- UNZ, Interchange Trailer

A FUNCTIONAL GROUP contains:
- UNG, Functional Group Header
- Messages of the same type
- UNE, Functional Group Trailer

A MESSAGE contains:
- UNH, Message Header
- Data Segments
- UNT, Message Trailer

A SEGMENT contains:
- A segment Tag
- Simple data elements or
- Composite data elements or both as applicable

A SEGMENT TAG contains:
- A segment code and, if explicit indication, repeating and nesting value(s)

A SIMPLE DATA ELEMENT contains:
- A single data element value

A COMPOSITE DATA ELEMENT contains:
- Component data elements

A COMPONENT DATA ELEMENT contains:
- A single data element value

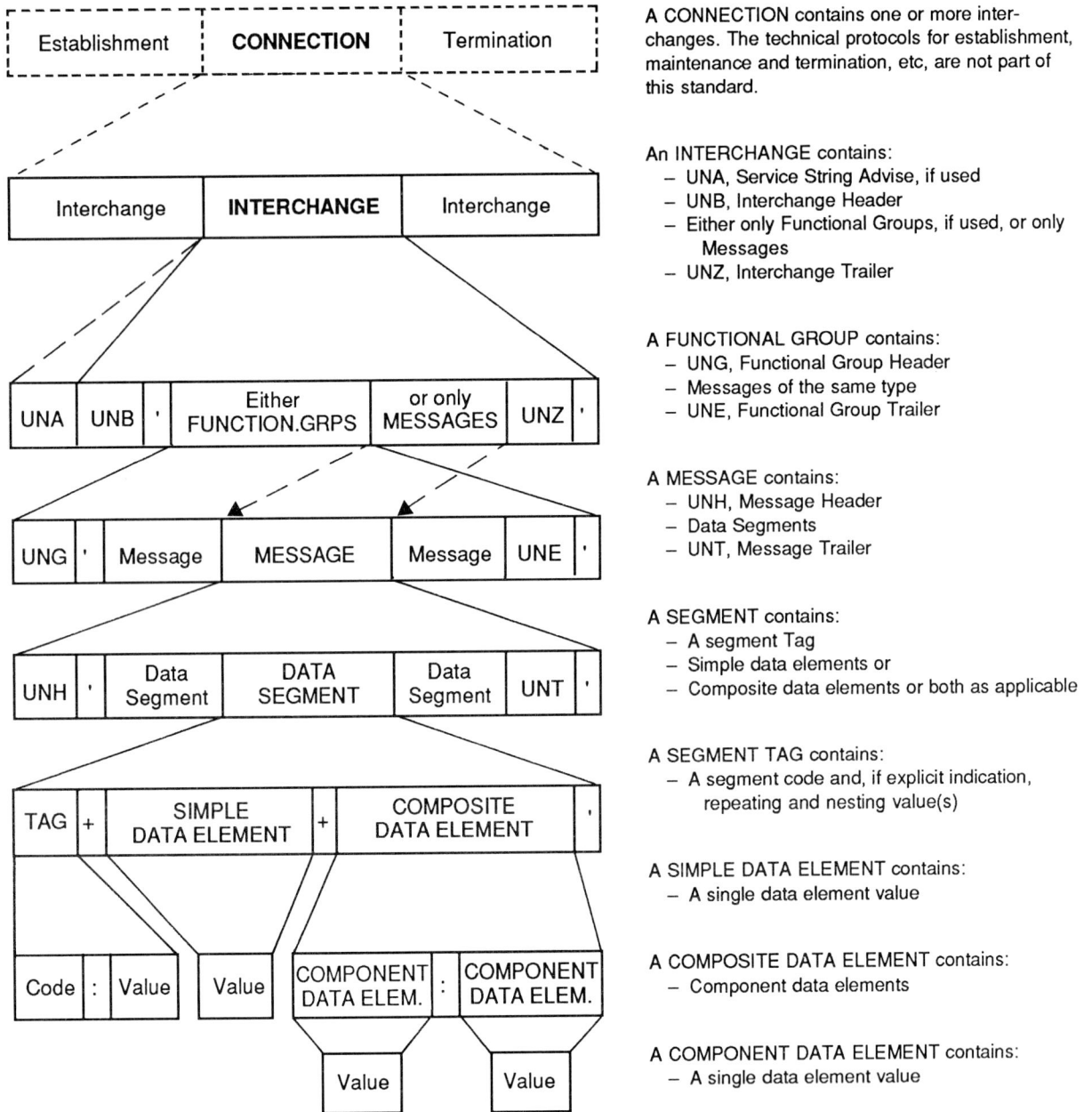

Figure 6.17 Extract of EDIFACT Standard

third of the major categories, communications.

EDI standards for data formats have been specifically designed so as to be independent of the communications medium which is being used. Hence, an interchange of trade information, structured in accordance with an EDI standard, can equally well be transferred by physical media (floppy disk or magnetic tape) or via a direct telecommunications link between computer systems. The particular communications option selected for an EDI application will depend entirely on the specific requirements for that application. Typical considerations are:

— speed of information delivery;

— security requirements;

— management requirements;

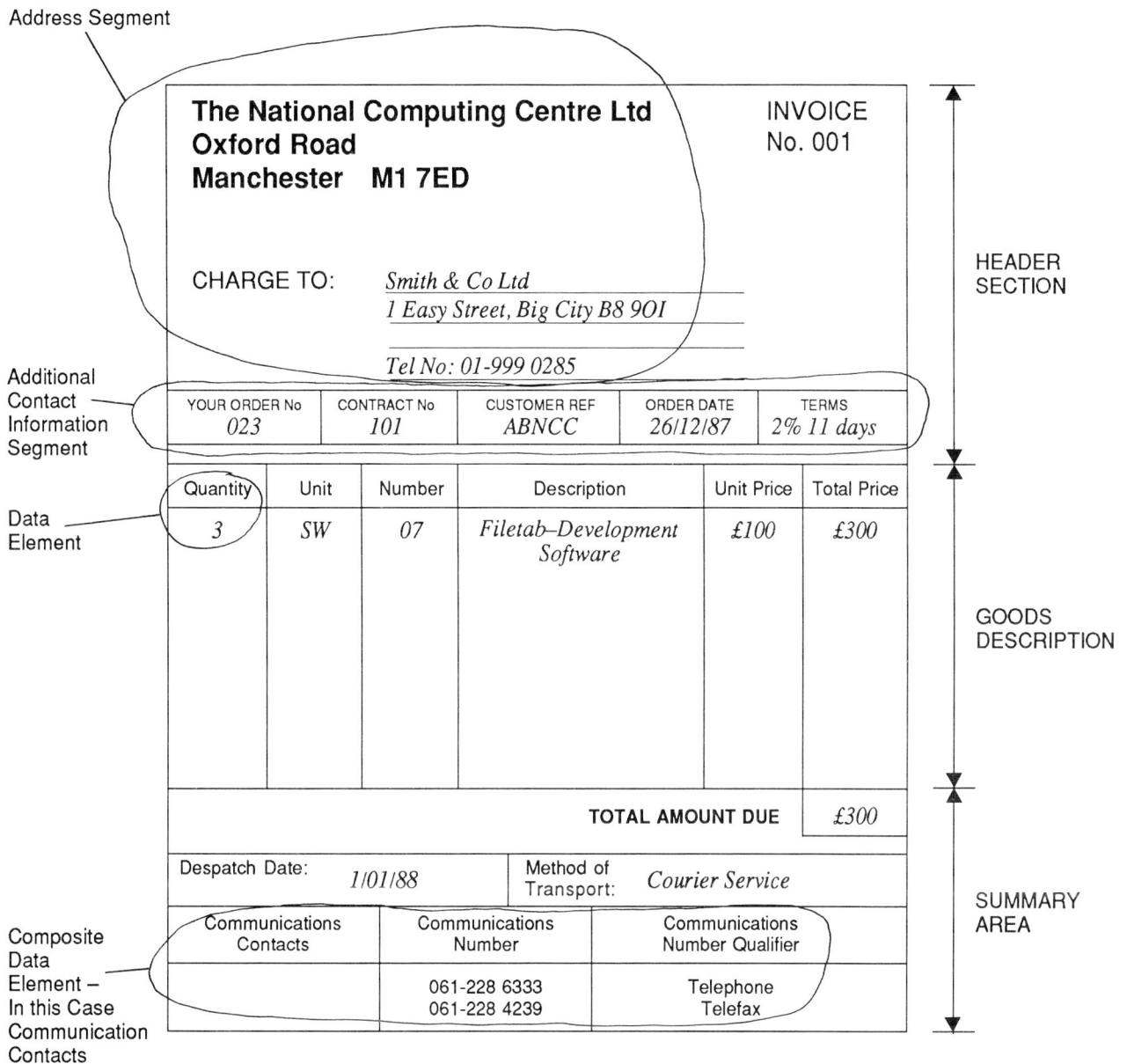

Figure 6.18   A Paper-based Document with its Elements and Sections Identified in EDIFACT Terminology

— decoupling from communications process;

— privacy requirements;

— availability;

— cost.

For the purpose of direct comparison with the use of X.400 for EDI it is only relevant to investigate the existing telecommunications options. These options are shown in Figure 6.19.

In Figure 6.19 the gap between the in-house user application and the various options at the lower levels indicates the degree of effort the user has to expend to get his EDI system running. If the basic bearer circuits are used then the user will be involved in a lot of additional implementation effort. If, however, the VADS option is taken, rather less customised implementation is required. The following sections discuss the options given in Figure 6.19 in more detail in order to highlight their strengths and weaknesses:

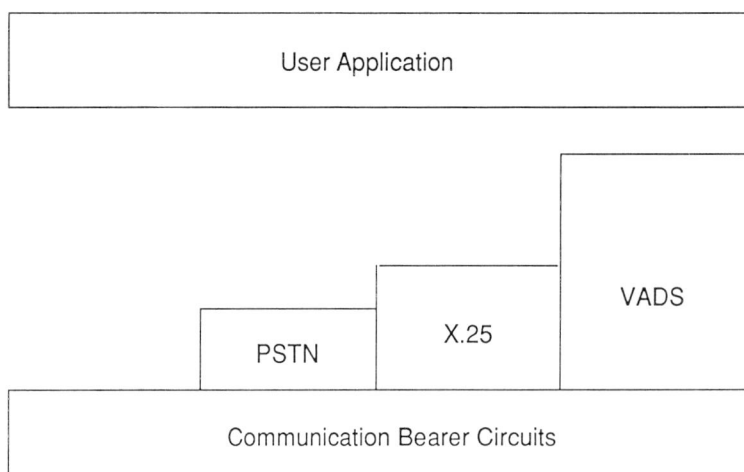

Figure 6.19   The Telecommunications Options for EDI

Leased Private Circuit Network

A network formed from leased bearer circuits can be attractive to certain organisations with special requirements. However, because of the effort implied by the establishment of each individual link this approach should only be considered when dealing with a relatively stable community. Common computer communication protocols need to be established for all links. With this approach, however, the user can build in as much or as little security, management and resilience as they require. In addition to this, such a network has good privacy associated with it which may also be important, eg SWIFT the banking network, where privacy is crucially important, use a leased line network.

Public Switched Telephone Network (PSTN)

Most EDI users are not in a very stable community and hence require some form of switched connection network. The PSTN is probably one which springs to mind first but it is not entirely suited to data transmission. Although the PSTN is gradually being migrated to full digital capability it is currently limited by the large amount of analogue equipment which is still in place. The bandwidth offered by the PSTN is approximately 3KHz which greatly limits the speed at which data can be transferred. In addition to this the use of analogue technology means that signals readily accumulate noise and hence can be grossly distorted when passing through the network. With these points in mind, however, the PSTN is widely available and does offer good flexibility. Again, with this approach, a common protocol is required between communicating systems along with a modem to convert the digital signals of a computer into an analogue form suitable for transmission in the PSTN.

Public Data Networks (X.25)

The X.25 public data network (PDNs) offer an international standardised service for data transmission. The international PDNs now provide coverage in most countries around the globe and because they have been developed to an international standard, most are interconnected. PDNs have been specifically designed for data transmission and so provide a reliable service with built-in error correction. There is still a requirement to implement a file transfer protocol to run on top of the communications functionality provided by the X.25 network.

Value Added and Data Services (VADS)

VADS offer what can effectively be called a clearing house service for EDI interchanges between trading partners. The service offered is basically store and retrieve. When a user wishes to send a message or interchange to a trading partner they merely have to send the data along with some addressing information to the VADS network service. At this stage the message will be placed in the mailbox of the intended recipient so that the recipient can retrieve it at a convenient time. The process works in exactly the same manner in the reverse direction as each subscriber to the service has a mailbox of their own.

The one advantage of this approach is that each user is effectively decoupled from the communications process, whereas with the other communications options described so far, both trading partners wishing to carry out EDI have to establish a communication session between themselves. This yields many problems, especially with large groups of trading partners in terms of having to co-ordinate times of availability at both ends. With a VADS network service the availability is generally 24 hours, hence sending data can easily be set up as an automatic process.

Protocol compatibility is no longer a problem with VADS because the service acts as a buffer between trading partners, so each organisation can use whatever protocol is suitable for their systems. All VADS support most of the popular communications protocols. The VADS providers have also gone some way to making EDI implementation a less traumatic process by offering support and consultancy services as well as standard software packages to act as the bridge between the customer's in-house application and the network service. Such software will also perform the conversion from in-house data formats to those of the popular EDI standards.

The major disadvantages of VADS is that a subscriber can only talk to other subscribers to the service. Indeed some organisations have found themselves in the unfortunate position of having to subscribe to more than one of these services in order to communicate with all their potential trading partners.

So those are the communications options which have been widely used for EDI to-date. It should be clear from the discussion that the VADS solution is the most suitable, mainly because they have been specifically tailored for this application. Indeed it is fair to say that VADS have been by far the most popular option for EDI so far. In the light of this it would be reasonable to ask if there is a need to use X.400 for EDI when the VADS cater so well for this application.

EDI OVER X.400

The clue to the need for X.400 lies in the weakness of VADS: their connectivity. Because X.400 services are based on international standards, connectivity is not a problem, indeed from an X.400 service providers point of view, links to the outside world will be a fair proportion of the 'added value' which their services can offer. With a complex infrastructure of worldwide X.400 links established, it is possible to conceive that such a service will provide a catalyst for the use of EDI in all trading operations. This scale of usage will imply a movement away from the current close trading partner relationships which are part of current EDI philosophy but this is likely to occur anyway when electronic trading gains wider acceptance. X.400 systems with their potential for multiple applications over the same connection will also provide a cost-effective solution for the EDI application.

Earlier chapters of this publication have already emphasised that the X.400 recommendations have been devised in such a way as to make the underlying Message Transfer (MT) service generic to all applications with the ability to transport any type of structured or unstructured data. This means that X.400 is eminently suitable to act as a platform for applications such as EDI. Early impressions would seem to indicate that X.400 will have a major impact on the emergence of EDI generally, via the interconnection of existing systems and services. Its ability to link systems will mean that existing investments are not wasted. Also because X.400 has been internationally agreed it will be the basis of a future proof communications medium.

The VADS suppliers have also been quick to realise the potential of X.400 as a medium for EDI and its possible impact on their existing businesses. However, if they adopt X.400 and use it to its best effect, they can quite easily profit as well. An X.400 interface will be another facility which they can offer their users. The VADS suppliers may additionally have a use for X.400 as a means of establishing links between their services, which will allow their previously separate user bases to gain access to each other. Evidence of the VADS suppliers interest in X.400 can be gained from a recent statement which INS Ltd, a joint ICL and GE IS venture, have supplied especially for this publication.

"We do believe that EDI could become a significant application for X.400. We have jointly agree with other suppliers in the industry to commission a study of the relationship between the standards in these two areas which will be completed early 1988.

We notice more and more in the invitations to tender received from European organisations the stipulation of X.400 as an optional method of conducting EDI and to this end have committed to providing such facilities in a number of cases to date, most notably the recent CEFIC initiative."

INS Ltd - 22 October 1987

The study into the relationship between X.400 and EDI has been launched by Vanguard, a UK government and industry sponsored initiative to promote UK industry awareness of VADS (INS are one of the group of industry bodies supporting Vanguard). Further information about the Vanguard EDI and X.400 study is contained in a later section. Also for further information about the CEFIC project refer to Chapter 5 and the section on GE IS.

THE TECHNICAL ASPECTS OF EDI OVER X.400

X.400 offers a store and forward message transfer base for certain types of EDI applications which are not of real-time nature. When an EDI transfer between organisations does not involve large amounts of data (ie more than a few Mbytes) X.400 is extremely suitable. In addition, the X.400 link will have a dual purpose, both as the corporate EDI and E-mail connection.

X.400 in its current form, ie the IPM and MT services, can only offer limited support for the EDI application. As the original user service, Interpersonal Messaging (IPM), was not designed to cope with EDI its service elements are generally only useful to the human user. While it is possible to send EDI information within the IPM structure it is not possible to identify it as such because EDI is not, nor should it be, a registered body part of IPM. In such a situation one has to dedicate a specific UA and O/R address just for EDI traffic. In this way it may be possible to prevent the situation where an unintelligible EDI data stream is delivered to an unsuspecting human user.

Another option which is possible with systems defined with the current 1984 recommendations is to bypass the normal route of a user service and make direct use of just the message transfer service as defined by the P1 protocol. P1 makes provision for designation of the content protocol being used, in the case of IPM this indication is P2. There is also an option to declare an undefined content type where private organisations can create their own content types. This undefined approach is usually termed the P0 approach and is eminently suitable for EDI. Using this method it is possible to make use of all the useful MT service elements while removing the chance of delivering a message to an IPM user.

In some cases existing EDI systems may be accessed via an X.400 service and this may require the provision of additional addressing attributes, eg a network address. In such cases it may be possible to use Domain Defined Attributes (DDAs) which are an X.400 feature to allow for the provision of extra addressing information. In this way data may be transferred to a specific department or system external to the messaging connection.

Because of the lack of suitability of the IPMS approach it is obvious that there is potential for the development of a special class of UA which can deal specifically with the requirements of EDI. It is possible that such an activity could be carried out on a private basis but this would obly be of use if a large group of trading partners were involved where many such EDI-UAs could communicate. The alternative is to look towards an internationally standardised solution whereby a new content protocol for EDI would be developed for inclusion in the X.400 recommendations. It is this latter area and others which are considered in the next section.

INTERNATIONAL STANDARDS ACTIVITY – X.400 AND EDI

The potential for EDI and more specifically EDIFACT traffic to be carried by X.400 systems has been quickly realised by a number of parties, and already work is underway on the development of the existing standards to allow them to cope more easily with the requirements of this application. For clarity, each of the developments is looked at separately:

— ANSI recently issued a proposal to the NIST group working on the US X.400 messaging profile, to carry out some modifications to allow EDI traffic to be handled. The modifications taken onboard by NIST are very simple and amount to little more than a ratification of the P0 or undefined P1 approach. This alteration is contained within an addendum to the NIST Implementor's Agreement for X.400, a copy of which is contained in Appendix 10 of this publication. These modifications basically specify that an EDI UA should support all the essential Message Transfer

(MT) service elements as these are considered to be the only ones which are of use for this application. The EDI message itself is then carried by the P1 envelope as an undefined content type, which leaves the option of carrying many types of EDI date rather than just constraining it to one specific type, such as ANSI X.12.

— The UK Government and Industry initiative, Vanguard, recently commissioned a technical study into EDI and X.400. This study used the new 1988 X.400 recommendations (see chapter 10) as its basis as there have been many new features defined which are applicable to the EDI application. The report has produced a series of technical recommendations for an X.400 EDI solution and these are now being tabled within the interim CCITT work on this topic.

— A draft question has already been proposed for the new CCITT 89/92 study period concerning the development of an EDI application protocol (Pedi ?) which would eventually become part of the X.400 recommendation series and hence should be adopted by ISO in their MOTIS work. An interim rapporteur has already been appointed and has commenced work on this topic during 88.

— The Swiss PTT have entered into a research project with the Swiss Institute (a University) to investigate the relationship between X.400 and EDI. They recently held a public seminar to publicise the results so far. Amongst the options outlined at the event were that the EDIFACT translation would be interfaced directly with the underlying P1 protocol (see Figure 6.20), hence making use of the basic transfer service and relying on the error messages that will eventually be included within EDIFACT. Although mainly concerned with EDIFACT data this approach is largely similar to that of the NIST X.400 working group. It is intended that the results of this work will be fed into the X.400 work during the CCITT 89/92 study period.

Message Size Restriction with X.400

In order that practical systems can be implemented, agreements have been formed by the standards profile groups (see Appendix 5) as to the minimum message sizes which should be supported. The sizes agreed so far are as follows:

- CEN/CENELEC = 2 Mbyte;

- NBS = 1 Mbyte (possibly extending to 2 Mbyte).

Although these place no constraints on normal text and office document transfer via X.400 systems the EDI situation is slightly different. Some EDI applications can be classified as bulk with batches of messages being anything from 10 to 20 Mbytes in length. In such situations X.400 is obviously not suitable and a bulk file transfer protocol such as FTAM would be more appropriate. However, for smaller scale applications X.400 is very suitable as most of its advantages would be of great use with the EDI application.

Conclusion

It is easy to see from the evidence shown that the use of X.400 systems for EDI traffic is already being

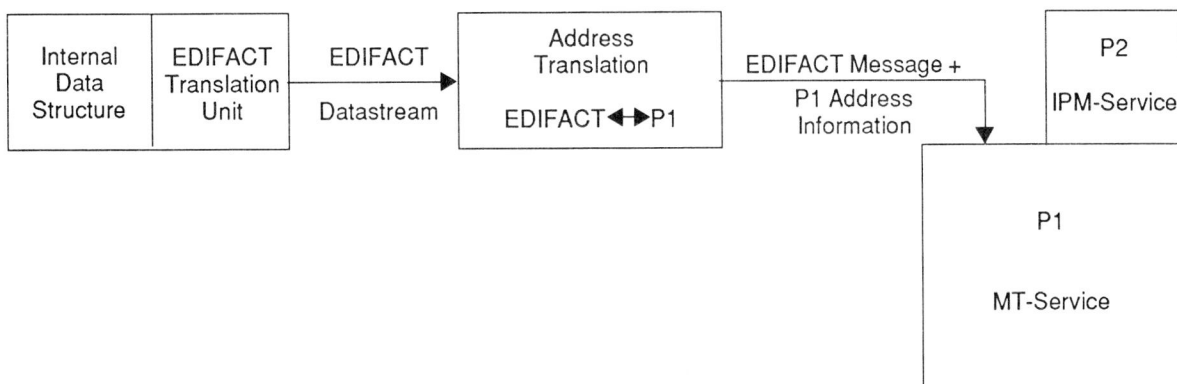

Figure 6.20   Direct EDIFACT Interface to the X.400 MT-Service

seriously considered by a number of parties. Hopefully such developments should bond these two technologies firmly together so that the benefits of each can be provided to the user in one package.

THE USE OF X.400 PROTOCOLS IN A DISTRIBUTED OFFICE SYSTEM

(*Author's Note:* Many thanks to Systems Designers plc and in particular Tony Whyman, upon who interesting work this part of the book is based.)

Before the LSI and VLSI (LSI=Large Scale Integration and VLSI=Very Large Scale Integration) revolution a powerful computer implied an expensive mainframe system. Organisations using such systems would have to base their networks around these large centralised devices with a distribution of relatively dumb terminals being the only means of access. Within the last decade, however, radical developments in the cost, size, storage capacity and processing power of computers have required a change of emphasis.

Moving away from centralised systems towards ones which are highly distributed marks the way to the future. It is now possible for departments, offices, and even individuals to have their own powerful processing tools. These new building blocks in distributed systems require efficient communication links in order to exploit their true usefulness, and Local Area Network (LAN) is having a substantial impact here. Along with the distributed system arrives the need for distributed applications, which will allow the network as a whole to interact in a coherent manner.

Distributed applications are not programs that exist in splendid isolation, accessing files here and terminals there. Rather they are a distributed set of functions: servers exporting services to client systems which require them. Generally, such applications are concerned only with information transfer, while activities such as information storage and how it will be presented on a workstation are purely a local function.

This section of the book will detail two examples of the use of the X.400 series of recommendations to implement functions within a distributed system. The examples are taken from projects currently being undertaken by Systems Designers, particularly in the area of office automation. In this field, the X.400 series of recommendations is seen to be of strategic importance as well as being an indication of future directions in OSI.

Of the examples discussed, the first represents a natural use of a message handling system (MHS) – in this case to implement a facility for the distribution of software – although the principles involved are clearly of wider application. The second is an example of the use of X.410 Remote Operations. These are used to provide the underlying communications support for a Diary Scheduler for diaries located on different systems.

Example 6.6   **Software Distribution**

In this example, a number of physically separate systems, each with local non-volatile storage, are attached to a network, constructed out of several interconnected local area networks (LANs), to form a connectionless internet. An electronic mail service, using X.400 protocols, is provided on this network, with each system supporting at least User Agent functionality, if not a Message Transfer Agent as well. Each system is installed with a basic set of systems software (including communications) but, thereafter, is expected to receive applications software and all systems software updates over the network from a central distribution point. Any updates will take place under strict change control in order to maintain a consistent set of software across all systems on the network.

The example network supports an organisation which, naturally, is structured into many departments. Each of these departments is an autonomous unit as far as configuration procedures are concerned, and these procedures are under the control of a departmental administrator.

As the internet is connectionless the mail service will be supported by a class 4 OSI transport service (see Appendix 3). The only implication of this is for interworking with a public MHS network which uses OSI transport class 0. However, it does not affect the X.400 protocols themselves.

The solution which SD have developed uses the X.411 message transfer (MT) service to pass messages between the following three distributed functions:

1 The Network Configuration Function. This operates at each departmental configuration point and

allows the various systems, their attributes, and the applications they support, to be defined by the departmental administrator. Once determined, the software requirements for each system are listed in a formatted X.400 message, which is then sent to the software distribution point to order the required software packages.

2 The Software Distribution Function. This operates at the software distribution point, and maintains the current software release and a database describing the systems and applications software requirements of each system on the network. This it updates from the 'orders' it receives from the configuration points. It is also responsible for distributing software, via X.400 protocols, to a newly installed system and scheduling the distribution of a new software release.

3 The Software Installation Function. This operates on every system on the network. It is responsible for receiving a new release of software and, on receiving the correct stimulus, for installing and using the new release.

In order to send messages to each other, these functions are recognised users of the mail service and have User Agents to interface them to the MT-service. The message flow between the above functions is illustrated in Figure 6.21, and reception of such a message is a stimulus to carry out some action dependent on the message contents.

Once a new system has been defined to the network configuration point (or an old one updated), the network configuration function determines its software requirements for both systems and applications software, formats the message ordering the required software sets, and sends this over the mail service to the software distribution point.

As currently defined, this message will be formatted as a text message according to X.420 P2 protocol for Interpersonal Mail (IPM). However, an alternative strategy, formatting the message as a locally defined bodypart with an abstract syntax specific to this application, would be equally valid.

The order will specify the applications and systems functions required but not the files containing the software, as the number and identity of the files may change between releases. On receipt of the order, the distribution function updates its database, determines the files that have to be sent, and sends them as one or more non-text X.400 interpersonal messages to the software installation process on the target system. When each one has been safely received and secured a confirmation is returned to the distribution point.

Once the complete set of software has been delivered, the system has to be reloaded with the newly delivered software. This act is termed 'commitment'. To ensure that this has been done, the commitment

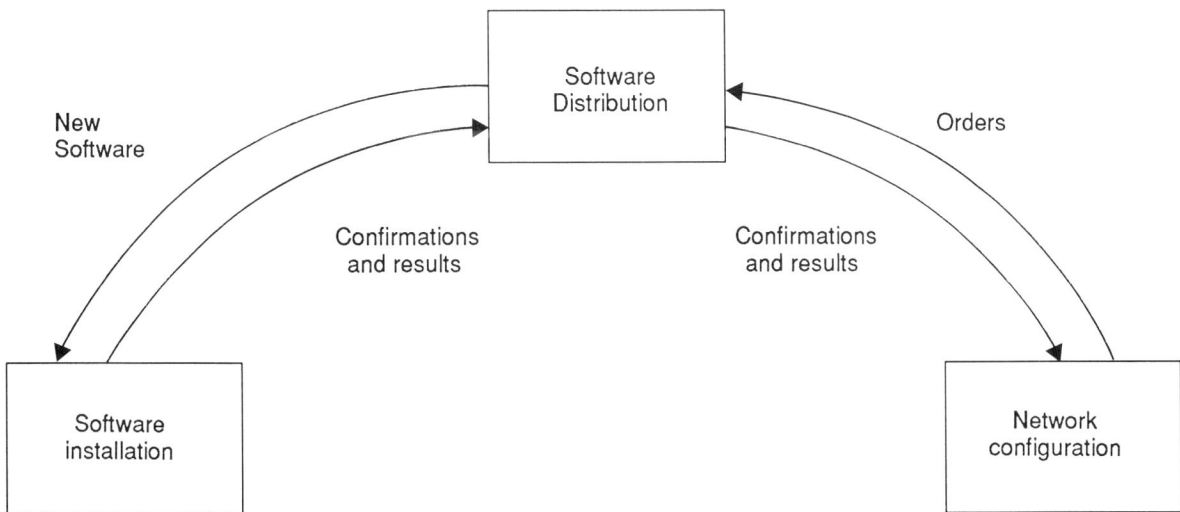

Figure 6.21 Message Flow Between Distributed Functions

must be performed under the control of the distribution function, which must be able to record the successful completion of the commitment—or failure—as the case may be.

However, the commitment cannot, in all cases, be actioned unilaterally from the distribution point. Because separate verification of the release may be required, or simply to ensure that the system is in a proper state to be reloaded, it will usually be necessary for an operator in charge of the software installation function to agree to the commitment.

It is thus necessary to define four further messages that pass between the distribution and installation functions to implement a two-phase commitment process:

1 Request that the installation function signal when it is ready to commit.

2 Inform the distribution function that the installation function is ready to commit.

3 Command the installation function to commit the new software and reload itself with the new software.

4 Report to the distribution function a successful commit and reload, or its failure.

The first message serves a preparatory function in that it informs the installation function, residing on the target system, that a software distribution is ready. The installation function informs the distribution function that is ready to accept and install the software with the second message. The third message initiates the installation of the software on the target system. The last message will be sent after the system has been reloaded.

On completion of a successful reload or its failure, the departmental administrator is informed of the success or failure of the software issue by a message sent to the configuration point.

A new release of software follows the same procedures except that the configuration function is not involved. Once a new release has been installed on the distribution point, the distribution function uses its database to determine distribution lists for each of the software sets and sends them as mail messages to the installation functions on each of the systems in the network. It relies on inherent functions in the mail service to optimise the distribution and the scheduling of the transfers, where possible.

When the distribution is complete, the advantage of the two-phase commitment may be seen. It is by no means guaranteed that two successive software releases may be compatible and it is necessary to bring all systems to a state where they are ready to commit and then commit them all simultaneously. The two-phase commit process achieves this.

The use of an X.400 messaging solution for this application may be contrasted with perhaps a more typical solution based on some file transfer protocol.

In essence, both achieve the same thing: the transfer of information – in this case software – from A to B. What is different about the use of a messaging protocol is that it is naturally distributive – ie in a single operation, the same information is sent efficiently to many destinations – and the act of delivering to a User Agent can be used to start some remote action. A file transfer protocol could be used, say, to transfer the software to each receiving system, but without building the equivalent of X.400's store and forward mechanism this could not be achieved with the same efficiency, across a large network.

Functionality on top of basic file transfer is required to control the commitment of the software. If this was done on a *connection oriented* point-to-point basis, say, then this is likely to tie up communications resources while waiting for the human interaction that is typically necessary to ensure that commitment of the new software does not interfere with other work. More likely, it would not be practicable, during a new software release, for the distribution point to have a concurrent connection with every system in the network. Connections would then have to be deliberately dropped by the distribution point after the first preparation message, and it would then have to start *polling* to determine when a system was ready to commit, with all the inefficiency that this would entail.

Again, the file transfer protocol would need to be given the equivalent functionality to X.400, if it was to give the same efficient and simple solution.

To summarise, the choice of an X.400 MT-service , as a bearer service for software distribution, has simplified the implementation and avoided the problems inherent in connection oriented protocols, across system breaks, or when delays are present due to a need to solicit human interaction. SD expect that an X.400 based solution will be both cheaper to implement and more efficient in operation.

Example 6.7 **Diary Scheduling**

In this example, a similar network to Example 6.6 is used. Electronic Diaries are implemented on several of the systems in the network, and a Diary Scheduler is required to schedule meetings by a co-ordinated update to the diaries belonging to potential attenders. Interactive response times (ie a few seconds) are required for the agreement of a provisional date, in order to minimise the length of time for which a diary may be locked. The meeting scheduling process is only completed when all propective attenders have indicated their agreement/refusal to attend, and the meeting has been confirmed or called off by the meeting's convener.

A model for such a scheduler is illustrated in Figure 6.22. Two functional units are identified:

1 A Schedule Master Function, which will typically run on the local system of a caller of the meeting, and is responsible for identifying a possible meeting date.

2 A Schedule Server Function, which is responsible for determining which dates the diary owner is available. It will be invoked on every system on which a diary is located that belongs to a propective meeting attender, including that of the caller of the meeting.

A Dairy Scheduler could, as in Example 6.6, be implemented over the MT-service, but to give interactive response times for scheduling the provisional date, which cannot be guaranteed by a mail service, the Remote Operations Service (ROS) of X.410 is used instead.

Remote Operations

Remote Operations form the underlying service for the X.411 P3 Submission Delivery Entity (SDE) protocol, which a remote User Agent uses to access its Message Transfer Agent, and, more generally, provide a framework for implementing a Remote Procedure Call mechanism.

(*Author's Note:* A procedure is a programming term which is employed to describe a series of regularly used instructions and/or calculations grouped together in order that they can be called by a name, from within a program. In calling such a procedure one would also pass some parameters upon which the

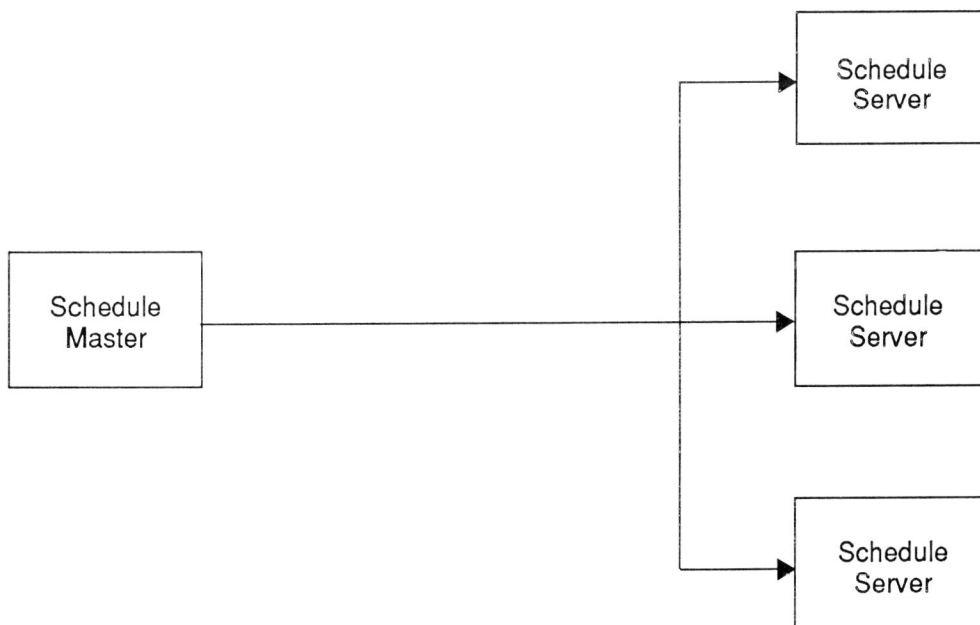

Figure 6.22 Model for Diary Scheduler

procedure would perform its operations. When the procedure function is complete the results of the operations are returned and normal program flow is resumed.)

Four message formats are defined in X.410 for remote operations, and these are transferred as discrete messages over the X.410 Reliable Transfer Service (RTS). They are:

Invoke – Which is used to invoke some operation (eg. a remote procedure) and to pass the operation's parameters;

Return Result – This is returned to the invoker of an operation upon successful completion along with any results from the operation;

Return Error – This signals an unsuccessful termination of the operation and passes back diagnostic data;

Reject – Is a 'catch-all' for responding to an unexpected or indecipherable message.

By defining a Remote Procedure Call as an operation, a remote operations service, when it is provided, can be used to allow rapid implementation of remote procedure calls. It is only necessary to provide, for each related set of procedures, a stub code which:

— locally provides the procedure interface; encodes the parameters; invokes the operation; and, on return, decodes the result;

— and, remotely, on invocation decodes the parameters; calls the actual procedure; encodes the result; and completes by returning the results or error status.

The architecture for Remote Procedure Calls is illustrated in Figure 6.23.

Applying Remote Operations To Diary Scheduling

If the interface between the schedule master function and its server is defined as a procedural one, then the first advantage that becomes apparent, if a remote procedure call mechanism is employed, is that when the server is invoked on the local system then it can be done by a direct procedure call. Indeed from the applications designer's viewpoint the communications aspect may be largely forgotten.

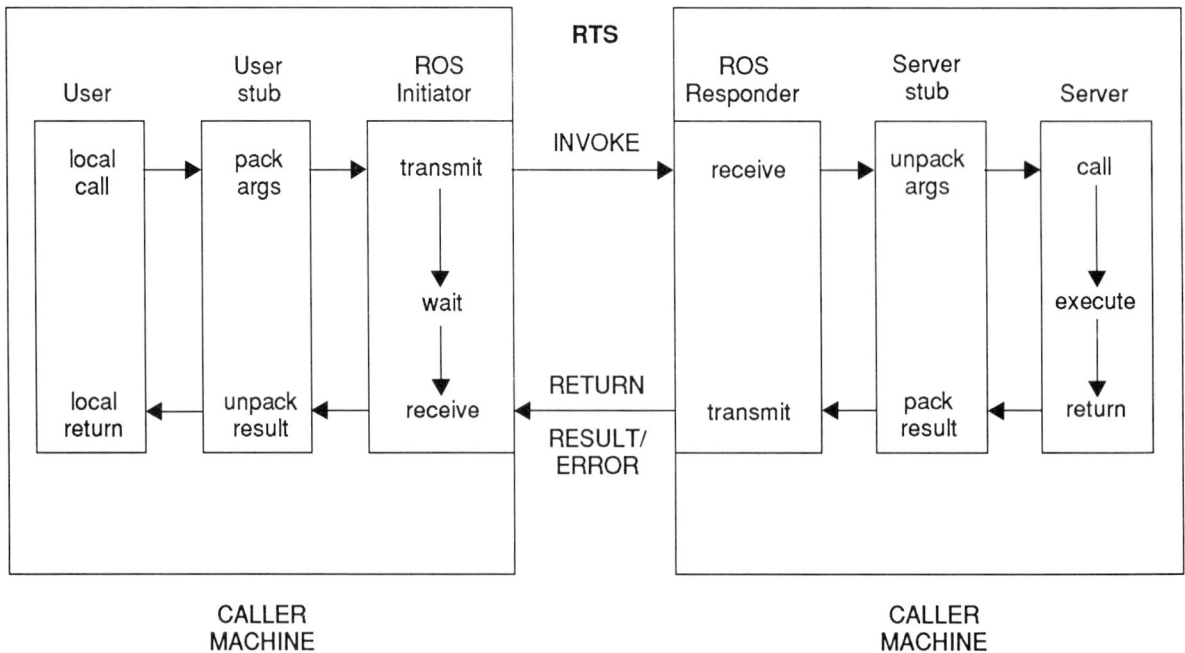

Figure 6.23   Remote Procedure Call Architecture

The following operations (remote procedures) are sufficient to schedule a provisional date:

1 Request, from a diary, a list of dates and times when it would be possible for the diary's owner to attend a meeting within a given period. Only free periods that were sufficiently long for the expected meeting length should be returned.

2 Insert a provisional entry in the diary for the meeting. Following the completion of the operation, the attention of the diary's owner is to be drawn to this entry at the earliest opportunity.

The use of these two operations would be: first, for the master function to obtain a list of all possible attender's free periods (probably including some room booking diary as well); and second, to determine a possible date for the meeting by finding the first free period common to all the lists. This provisional meeting date would then be entered in each diary by the second operation.

However, the two above operations must occur in the order presented, with the diary locked between their invocations. If not, then double booking could occur through concurrent usage of the diary scheduler.

Such a requirement naturally brings to mind the OSI Application Control Service Elements (ACSE) for Commitment, Concurrency and Recovery (CCR—ISO 8650/3) service (see Appendix 3). This is one of the basic 'tool kits' provided by OSI upon which applications are based. CCR provides control over the 'atomic actions' or indivisible actions of connection. With basic protocols used for information transfer material will have to be duplicated, ie retransmitted, if no final acknowledgement is received. CCR handshakes overcome this problem by ensuring that in the event of failure an accurate idea of the status of the tramsmission, eg about to commence, in the middle or finished. In this way information only has to be retransmitted when it has been lost. The handshakes defined within CCR are as follows:

C-Begin – initiates an atomic action;

C-Prepare – optional, used in circumstances were the end of an atomic action is not clear to the responder;

C-Ready – the responder offers to commit at the end of the atomic action;

C-Refuse – a refusal to commit, issued at any time before the issue of C-Ready;

C-Commit – the order to commit issued after a ready to commit has been received;

C-Rollback – issued at any time before the commit order to rollback the operation.

The C-BEGIN and C-PREPARE commands can bracket the two diary operations, defined earlier, respectively to lock and unlock the diary and, subsequently, C-READY can be used to imply agreement by the diary's owner. Conversely, C-REFUSE can imply that the diary's owner will not attend, and C-COMMIT and C-ROLLBACK can then be used to confirm the meeting or call it off as the case may be.

As a presentation layer connection supports the Reliable Transfer Service used by Remote Operations, there would appear to be scope for combining the CCR protocol of ISO 8650/3 with remote operations. However, the above sequence of events demands human interaction. There may be a period of days between making a provisional entry and its subsequent acceptance or rejection, and it is certainly not desirable, or even possible if the systems are switched off at night, for connections to be maintained throughout this period.

If the standard CCR were used the problems of dropping and re-establishing associations outlined earlier would be encountered.

This can be avoided by an alternative solution which involves adding C-BEGIN and C-PREPARE semantics to the two diary operations outlined previously, but then using X.400 Message Transfer Service to convey the subsequent CCR messages corresponding to agreement or refusal to attend the meeting, and to confirm or call off the meeting.

Such a hybrid approach gives the response times desired for getting a possible meeting date, but then simply enables the subsequent actions to be achieved without the complications of managing connections.

Conclusion

Perhaps the most important observation to make from the above examples is that use of X.400 services as bearer services has enabled application associations to be decoupled from the underlying connections:

in this context the connection being used in the sense of virtual circuit. In each case the association exists until the required action has been completed. During this time several connections may have been opened and closed, but the applications have concerned themselves only with message exchange, while the underlying service provider determines when connections are active.

The advantage of messaging techniques can be most graphically seen in the case of a directory service. This is characterised by a large number of potential service users that make infrequent use of the service. In most cases it is impractical to maintain a permanent connection between the directory service and each of its users, and setting up a connection for each access is expensive and time consuming. Accessing the directory by discrete messages minimises the communications overheads at the applications level, expedites the access, and allows the directory to be implemented as no more than a single-state machine.

A similar result is true for Systems Management, where the Management Station potentially has to be in contact with all network entities in its domain.

The service offered by the X.411 MT-service is simply a connectionless, albeit reliable, message transfer service. However, the underlying protocol used means that it is only at its best when message exchange is frequent or a large message is being transferred. When message exchange is relatively infrequent and the messages are short, as is typical with remote operations, it is far from optimal and it is thus worth investigating the possibility of using alternative 'lightweight' bearer services, offering a similar connectionless service, but with only minimal enhancement of the underlying network service, such as that provided by the connectionless internet protocol.

IMAGE TRANSFER OVER X.400 SYSTEMS

Currently it is estimated that about 95 per cent of the information held by commercial and governmental organisations within an office environment is unstructured, held in image form on paper documents. Whilst it is in this state, information cannot be searched, retrieved, edited or transmitted by computer-based systems. A further 4 per cent of all information handled by organisations is held in unstructured form, stored on microfilm or microform storage media. Computer assisted retrieval (CAR) techniques may be used, whereby indexing information is stored on and searched by a computer.

Technological developments are making it practical and economically feasible for more and more organisations to capture the information in documents electronically and store it in digital form, eg on WORM (Write Once Read Many Times) optical disks. This 'image' information can then be accessed and managed, as data files and text files within office automation and data processing environments and—where the application demands—integrated with these environments.

WHAT IS A DOCUMENT IMAGING SYSTEM?

A document imaging system (DIS) allows users to scan documents, display the images, and print them. Typical resolutions are respectively 100, 200 and 300 dpi (dots per inch), with higher resolutions provided if the application justifies the additional costs and higher usage times that may be incurred. The DIS may also allow a user to enhance the images and – where the application calls for more than filing, retrieval and printing – modify images. More sophisticated DIS will allow image, text and data files to be stored on the same storage medium, will allow mixed-mode documents (with image, text, data and, perhaps, voice) to be generated and processed, and will allow the textual content of an image to be recognised by optical character recognition (OCR) techniques and coded into normal ASCII based format.

DIS are not intended to be standalone devices, rather they have a useful purpose to serve in the integration of normal text-based document environments and those of image. To this end communications between DIS and other office equipment are of significant importance. Some systems already allow the user to transfer images via LAN, but this can be considered as only the first stage. Communications between LANs and to many other systems external to an organisation are also important. X.400 message handling systems can offer an internationally standardised solution to wide area DIS communications. Because X.400 can offer an all-purpose message transfer service for any kind of structured or unstructured data it is an ideal platform for image or composite image document transfer. Ultimately it is possible to conceive of image documents being transferred outside the office environment to all parts of the globe via a worldwide X.400 messaging infrastructure.

FUNCTIONS OF A DOCUMENT IMAGING SYSTEM

Image Capture

The component which allows a DIS to capture a paper-based image is called a scanner. The scanner scans the document and encodes its black and white elements as a data stream, which is exactly the same technology as the facsimile machine. The resultant data stream is then buffered, displayed, compressed (a technique for reducing the amount of data to be stored) and output to magnetic disks for temporary storage and indexing before final storage on an optical disk. Compression ratios of about 10:1 are often achieved: thus an A4 page produces (at 200 dpi scan resolution) about 4 Mbits (uncompressed) and 50 Kbytes after compression. This compression technique increases the number of documents which can be held within a given storage area, reduces the time taken to write an image/to read an image from a disk, and also reduces the transmission time over a LAN. Compression is also very useful when transferring a document image via an X.400 service (such as BT's Gold 400) because charges are based on message size.

Indexing

A document image stored within a DIS will need to be retrieved from amongst the other documents held on the storage medium, this requires an indexing system to be employed. The sophistication of the indexing system will be appropriate for the specific user application.

Storage

The permanent storage medium for compressed, digitised images is usually WORM optical disks at present, although magnetic disk may be used for very active files. For PC-based DIS, 5.25 inch WORM drives, with up to 450 MBytes of storage per side may be used. Other systems, based on microcomputers, may use 12 inch worm drives, with a capacity of up to 1.8 GB per side of WORM disk (in excess of 50,000 A4 document images per disk).

Retrieval

A user's retrieval station may be linked to a system controller (or host) which holds the document image index database and, through that, to the image storage subsystem. The retrieval process will be similar to that for textual document retrieval and may be menu-driven.

X.400 FOR IMAGE TRANSFER

It is possible to transfer document images via X.400 as interpersonal message (IPM) body parts. The IPM service defines an electronic mail service for the exchange of messages between human users. In the case of DIS users this would be the basis for their electronic mail facility. There are a number of standard body parts defined for an IPM, one of these is Grp 3 facsimile (see Chapters 2 and 3). Using the X.400 IPM service it is then feasible to transfer document images encoded in accordance with Grp 3 facsimile between DIS.

In order to allow more flexibility it is reasonable to transfer any type of image or composite text/image documents via the X.400 IPM service by allowing them to assume the status of an 'undefined' body type. This approach would obviously require the user to confirm that the recipient's system has the capability to handle such a document in advance of sending ot. The purpose of the standard 'body part' indications is to allow systems to automatically interrogate the message header to assess whether they are capable of handling the contents of the message.

In the future, as composite text and image documents become more common, DIS systems will have to include facilities for applying a standardised document architecture such as ISOs Office Document Architecture (ODA - see this chapter). This will allow layout and logical structure information about the document (as specified by the originator during its creation) to be transferred along with it. Such an approach will not present any difficulties when considering transfer via an X.400 system.

XIONICS AND X.400

After considering the technical feasibility of image transfer via X.400 it is interesting to look at the way in which one major DIS vendor intends to integrate X.400 within their systems. Xionics, a leading company in the provision of DIS, have provided the following statement detailing their approach to image transfer via X.400.

(*Author's Note:* Many thanks to Xionics, in particular John Cummins, for their contribution to this publication.)

About Xionics Document Image Processing Systems (DIP-X)

From a background of networking and office automation implementations, Xionics have pioneered the exploitation of PC based, low cost Document Image Processing (DIP) systems since identifying it as a mass market opportunity in 1986.

XIONICS have consequently designed a dedicated PC image card and bundled software termed 'DIP-X' which give a standard IBM PC powerful and comprehensive DIP capabilities. A novel and unique feature of DIP-X is its ability to be completely transparently driven by other software through its Command Line Interface (CLI). This enables it to be closely integrated with existing office automation software giving it an extra imaging dimension.

XIONICS can supply a wide range of systems and support packages ranging from the supply of the DIP-X card and software through to standalone and networked DIP-X stations. A software consultancy service is available so that complete turnkey solutions can be provided to clients which emulate their existing paper-based office procedures.

Investment levels required to join the document image processing movement range from £1,500 for the DIP-X card/software to c £20,000 for a turnkey standalone DIP-X workstation.

XIONICS and Standard Groups

XIONICS is an active member of messaging implementation groups, in particular IGOSIS X.400, IGOSIS ODA and British Office Technology Manufacturers Alliance (BOTMA) X.400 support groups.

We are planning to be involved in conformance testing and will participate in a rigorous interworking/ evalutation scheme of our customers who will be adopting X.400 within their corporations.

XIONICS X.400 Product Plans

XIONICS is commited to offering existing users of XIONICS office automation systems X.400 gateway facilities to ADministration Management Domains (ADMDs) with the ability to form corporate wide PRivate Management Domains (PRMDs).

These gateways will also be offered as low cost nodes suitable for networking via Ethernet and IBM Token Ring Local Area Networks (LANs). The nodes will consist of Intel 80386 processors with UNIX V.3 operating system and high performance integral X.25 communications. UA/MTA facilities will be supported.

The O/R naming system provided (Form 1, variant 1) is designed to allow interworking with BT's ADMD.

It is planned to converge with CEN/CENELEC Functional Standard for X.400 and support BT's ONA and IDEM (except for certain security aspects).

The user interface will be based on XIONIC's existing communications sub-system carefully modified to take into account customer responses and results of extensive X.400 simulations.

Body parts supported will be initially restricted to text (IA5) and 'Unidentified body part'. 'Unidentified body part' will be used in Document Image Processing (DIP) systems to transport CCITT G3 and G4 facsimile compressed images between nodes in a MHS domain. G4 facsimile compression is preferred as it produces the optimum level of image compression (typically 20:1). An A4 document scanned at 300 dpi resolution can be typically compressed to approx 50 Kbytes per page using this technique 1 Mbyte uncompressed.

In this context, X.400 is considered to offer a vital strategic role in providing a transfer service to support a 'raft' of integrated text and image applications.

Document Structures

Until ODA emerges as a complete and robust architecture, it is unlikely to be the first document structure to be carried by the Message Transfer (MT) Service.

Market and customer pressure will invariably dictate that IBM's DCA will be the first such structure carried. This whole area is full of complex and fast moving technical/commercial issues and requires very responsive tracking.

XIONICS strategy in these areas will involve first offering these standards as interconnection services for text and image application systems whilst evaluating them as a suitable basis for future turnkey systems.

THE MHS PLATFORM AS A MEANS OF ACCESS TO COMMERCIAL SERVICES

Within many countries X.400 based message handling services are being developed by the major administrations and service providers. It is these services which will form the basic interconnection platforms within their countries of origin. In addition they will also provide international gateways, access to telex and other telematic services. In short an X.400 service will be the basis for the integration of existing forms of communication with new ones and the means by which a country's scope for electronic communication will be expanded to encompass all parts of the globe.

The potential for easy electronic communications offered by these services will mean that eventually they will take substantial business away from the postal services. This implies that companies will take a radically new view of the ways in which they offer their services. Where information transfer was via the postal service it may migrate to this new electronic media. Companies will begin to carry their X.400 addresses within adverts and on business cards in the way that they currently carry a telephone number. Certain types of service offered by companies may also migrate to a completely electronic means of operation, possible examples of which are discussed below.

In general though, X.400 is likely to have a major effect on the way organisations carry out many of their everyday business operations. Eventually it would seem that X.400 will become an indispensable tool for industry.

EXAMPLES OF SERVICES WHICH COULD BE OFFERED VIA X.400

Document Format Conversions

Many companies offer disk conversion facilities which amount to a conversion from one word processor format to that of another. Currently this is achieved by physically taking a disk to the company for conversion. In the future this could be achieved more efficiently by transferring the material via X.400 directly to the company who would carry out the conversion and return it electronically, again as an X.400 message.

Document Translation

Document translation facilities are provided by some companies who translate technical documentation from one language to another. In this case the same scenario as above applies, the client could transfer the original to the company and receive the translated document over X.400.

Legal Services

An organisation formulating a contract may wish to have it checked to ensure that its legal position is correctly defined. In such a situation the draft of a contract could be sent electronically over X.400 to a solicitor for checking and returned with corrections made. This may be an iterative process which could be easily and quickly achieved with X.400.

Information Services

Information services could receive requests for information over X.400 and return it by the same means.

Credit Checking Services

Requests to verify customers' credit ratings could also be quickly and efficiently carried out via X.400.

7 The Benefits of Using X.400 Message Handling Systems

Some of the benefits of using the X.400 recommendations as a basis for electronic messaging systems have already been stressed in earlier chapters of this book. It is, however, important to pull all of these together in a chapter which will provide a clearer insight into the range of advantages of this approach. This information will undoubtedly be of use to organisations attempting to find the required justification for expenditure on X.400 products as it may highlight certain implications which they are not aware of. Individuals who are doubtful of the worth of such systems may also be swayed in the light of some of the potential benefits listed.

Because of the lack of real user experience with X.400, at the time of writing it is not possible to refer to 'glowing' case studies which show the excellence of this technology. Instead, reliance must be placed on the potential which this new solution can offer in comparison with existing systems (see Chapter 2). The rationale will be to list the potential areas of benefit to be achieved by using X.400 systems and, in years to come as practical experience is gained, these should become apparent in reality.

For clarity the benefits presented here are shown under two broad headings:

— benefits to the user;
— benefits to the suppliers.

These headings can be further subdivided into more obvious categories:

— benefits to the user:
  • at an organisational level,
  • and at the end-user level;
— benefits to the suppliers:
  • the service providers,
  • and the equipment vendors.

With this structure the reasons for the use of X.400 Message Handling Systems (MHSs) and the suppliers adoption/commitment should become clear.

BENEFITS TO THE USER

Potential benefits to the users of X.400 systems are wide ranging and encompass far more than would be expected from any other existing form of electronic messaging media. Part of this stems from the overriding flexibility which is built into the original recommendations. This flexibility allows X.400 systems to act as a platform for many different applications. The case is helped still further because this is an internationally standardised solution which can be implemented by all countries with security in the knowledge that it is not going to change over night.

Some of the benefits to the user discussed within this chapter will become apparent right from the outset with an X.400 implementation while others may take a little longer. This is because some of the benefits will only emerge out of the establishment of a worldwide X.400 based messaging infrastructure which will undoubtedly take some years to develop. Other benefits will only become a reality when an organisation as a whole has migrated to this technology.

THE ORGANISATION

Migration towards X.400 within an organisation can only be achieved via high-level commitment. This kind of commitment can only be reached when the positive effects of such a policy have been clearly identified. It is thus vitally important to look at the benefits of X.400 at an organisation-wide level.

In this context the following benefits have been identified which are likely to be obtained by the organisation, as a whole, via the adoption of X.400:

Strategic Issues

Maintaining a market position. An important aim of most businesses is to establish a good position in the market. This can be done in many ways, one of which is to obtain a reputation for fast, reliable service to customers. In this context an efficient internal and external electronic messaging service is crucial. In the case of an X.400 system it can be used to access many existing forms of messaging media. An X.400 link offers the most suitable means of accessing all the internal branches within an organisation and a suitable basis for direct customer communication. In this way a faster, more efficient customer service would be easy to achieve especially when compared to the outdated postal services.

Achieving competitive edge. This is another, similar, reason which would be a valid one for adopting an X.400 solution. If the use of X.400 can produce gains in efficiency and in some cases reductions in cost, these savings can be passed on to the customer in the form of reduced costs or expanded services for the same costs. Staying ahead is a constant battle though, and it is not sufficient to improve efficiency and then sit back and hope to maintain the competitive edge. If competitors choose to adopt the same approach they will soon catch up and possibly take over the leading position. To remain competitive a company must develop and integrate the technology in the best way possible. Because X.400 is a future proof solution and can act as a basis for many different applications, it can grow with an organisation to meet its needs now and in the future.

Enhanced image. In a business where it is important to be seen to be keeping up to date with technology, it may be possible to enhance the image of a company by using X.400 electronic messaging facilities. With X.400 being based upon the principles of Open Systems Interconnection (OSI) it can be perceived instantly that the organisation is looking towards the longer term benefits to be gained by the use of technology. It is also possible that a company may wish to publicise its use of messaging handling via case study material which can be made available to customers. The fact that potential customers could communicate with the organisation via a whole range of electronic media must only serve to impress them.

Improved Communications

Information access. X.400 can have an important role to play in the integration of many existing communication and information systems within an organisation. Also easier external communications will allow access to many more sources of information. This means that users of the system will have better access to more sources of information and policy makers should be able to make decisions based upon a better selection of high quality information. The result of this scenario must surely be improved productivity from the existence of an X.400 system.

Quality of communications. Both internal and external communications will be vastly improved with the widespread adoption of X.400. Within a short time period it should be possible to see an extensive national and international messaging infrastructure developing. This, combined with large-scale penetration of internal X.400 communications within companies, will serve to lower communication barriers which have existed for so long. This improved quality of communications will benefit all tiers of the organisation involved in accessing and disseminating information.

System integration and interconnection. Because X.400 is an internationally standardised solution it can be used as a general means for the interconnection and integration of systems. For the organisation this means that X.400 can be used as the 'glue' with which to bring together their existing standalone means of communications. Taking the example of integrated office systems, these generally offer useful mail facilities to small pockets of users within an organisation. Utilising X.400 gateways between these allows users of each of the systems to exchange messages, thereby establishing closer working relationships between the groups within an organisation.

Purchasing Equipment

Cost reductions. Because X.400 can offer a single solution to a number of, if not all, a company's messaging requirements, there are potential cost savings to be made. This streamlining of communications means that an organisation need only subscribe to one X.400 service rather than a number of others. Although this service in itself may be more expensive than any of the previous individual services, it is unlikely that it would exceed the total bill for communications, hence a cost saving in real terms. These kind of cost savings can only be achieved when the full potential of X.400 is being used, utilising one link for many applications.

Protection of investments. X.400 functionality can be added to most existing hardware via a software package which will either allow 'native-mode' or gateway operation. In this way X.400 can be used to protect existing hardware, and in some cases software as well, investments. Also, the addition of X.400 functionality adds real 'value' to existing systems by allowing them to address a far wider user base than before.

Future proof solutions. X.400 is an internationally agreed solution which is intended to be the basis of systems for some considerable time. Any future updates of the recommendations will provide backward compatibility paths so as to protect existing investments. In this way systems will become future proof as they will always be able to communicate with other X.400 systems regardless of age and model.

No more special gateways. X.400 should become the commonly accepted means for systems interconnection and integration. In this way the need for specialised gateways, between proprietary systems architectures, will be removed. In order to interconnect with other systems a manufacturer need only supply a standard X.400 gateway which will meet all the needs of the user. This in itself suggests substantial savings both in the effort and cost required to produce a suitable gateway function.

Freedom of choice. Freedom to choose equipment on the basis of merit rather than on the basis of whether it will interconnect/integrate with existing purchases is an added benefit of X.400. Because all the equipment from the vendors should conform to the same standard, organisations will no longer have to be constrained in their range of options.

Equal footing for suppliers. Suppliers can be dealt with on a more equal footing because the constraints of a particular proprietary system which are currently tolerated can be avoided. The selection of another supplier's equipment will no longer carry with it associated overheads which relate to compatibility. Easy X.400 interworking will reduce the dependence upon one supplier.

Increased competition between suppliers. Because suppliers will have to compete on an equal footing this should mean that equipment prices will start to fall while the quality will increase. This scenario has been shown to be true in many other markets where standards have applied - the de facto personal computer standard of the IBM PC for example.

X.400 Added Value

Internationally standardised solution. X.400 is the first internationally agreed standard for computer-based message handling. This means that its use will expose organisations to benefits which are not apparent with existing forms of messaging media. All round the world organisations, service providers and authorities are starting to adopt X.400 as the basis of their future messaging requirements (see Chapter 5), and these developments will lead to the establishment of a worldwide messaging infrastructure. This will mean that organisations adopting this solution will be able to address, via electronic communications, a far wider marketplace than with any other system.

A flexible application base. It was intended from the outset of their development that the X.400 recommendations would serve as a flexible application base for industry specific application in addition to simply electronic mail. A brief insight into this flexibility was given in the previous chapter where a number of potential applications were highlighted. For these reasons an organisation's X.400 links must be considered to offer more than merely electronic mail if their true worth is to be achieved. No other form of messaging media can offer this sort of adaptability.

THE END-USER

So far, the benefits of X.400 which can be perceived at an organisational level have been discussed, but there are those benefits which will become apparent to the actual users of systems. These will mainly be in terms of improved ease of use and access. But in time, as the usefulness of X.400 comes to fruition, it will become an indispensable tool in the everyday work of individuals. The advantages of X.400 to the end-user are:

Easy access. An X.400 link will give access to many other systems and services, such as telex, other telematic services and computer-based messaging systems/services. For the user this will mean that all messages can be sent from one terminal located on their desk. No longer will it be necessary to learn how to access many different systems, generally achieved with different terminals within separate offices. X.400 will give the user unrivalled easy access to all parts of the globe from a single terminal on their desk.

Information access. Because X.400 is a means of achieving both easy internal and external interconnection and communications with other systems and individuals it will yield access to much information which was previously either impossible or very difficult to get hold of. This improved information access leads to higher personal productivity, less wasted time, and more well-informed decision making.

Potential for many applications. Message handling systems have been defined in such a way that they may act as a basis for many applications, other than just electronic mail. Ultimately this will mean that users will be able to operate many applications form a single terminal.

Advantages over the telephone. These are benefits which can be incurred with the use of any electronic text communication service, but in the case of X.400 its impact will be substantially wider and its benefits more apparent:

— easy access across time zones;

— no need for 'telephone-tag' scenario;

— hard copy of a message removes the need for rough, sometimes inaccurate, telephone notes;

— text-based messages are more accurate and well defined;

— fear of telephone communications is eliminated.

Less reliance on slow postal services. In comparison with other forms of electronic messaging X.400 will provide access to a far wider audience. This will mean that users will place much less reliance on outdated and slow postal methods as their means of everyday communication.

BENEFITS TO THE SUPPLIERS

In order for message handling products and services to have appeared, as they are doing now, there must be benefits which the suppliers feel offer sufficient incentive. This section will attempt to identify these. For clarity it is divided under two headings:

— service providers;

— equipment vendors.

THE SERVICE PROVIDERS

The service providers are those that will provide message handling platforms which private domains will connect to. These platforms will provide features such as connection to other private domains, access to other national and international messaging platforms, connection to non-X.400 electronic mail services. gateways to telex and telematic services.

The advantages of adopting X.400 for a service provider are as follows:

Internationally standardised. X.400 offers an internationally standardised solution to the electronic messaging requirements of users. This is not a technological 'blind-alley', and systems based on X.400 will probably be around until well into the next century. Any development work on the current recommendations will ensure backward compatibility so as to protect existing equipment investments. The potential for a wide range of national and international interconnections to existing user bases means that it is substantially easier to achieve a 'critical-mass' of users necessary to provide a viable service.

Standard software. Standard message handling software is now available upon which it is possible to base a service. This implies a substantial reduction in development costs and the time scales with which a service can be brought 'on-line'.

Better User Service. Utilising X.400 as the basis for a service can offer substantial benefits which will make it attractive to potential users. International interconnection, access to telex and telematic services are all highly desirable features for users. In addition to this, it is also possible to exploit the potential of the service for the operation of applications other than electronic mail, such as Electronic Data Interchange (EDI).

User Package Availability. Already many X.400 packages are available for all types of computer hardware, such as mainframes, minis, local area networks (LANs) and micros. These will form the means of access to an X.400 service and so there are no worries about developing special access packages for this purpose. An additional point is that with the widespread use of X.400 within office environments there will be a large market to which the service can address itself.

THE EQUIPMENT VENDORS

The equipment vendors are the group who develop and sell the X.400 products for user systems. It is obvious that without commitment from this sector the widespread usage of message handling would never come to fruition. These companies are becoming heavily involved in X.400 developments and many can already either offer products or show their resolution by quoting dates for the launch of products (see Appendix 8).

The reasons for the widespread adoption of the X.400 recommendations by equipment vendors are as follows:

Large marketplace. Because all users will eventually require access to the benefits of message handling the vendor will be able to sell to a far wider marketplace. In the past this may have been restricted to either new users or those with existing investments in a certain range of products. Since many of the telephone administrations and service providers are looking seriously at X.400 there is also a lucrative market for the sale of complete systems.

Increased competition. Most equipment vendors will relish the idea of competition on an even basis. When products are developed to an international standard this situation exists. Products will have to compete purely on their excellence, technical or otherwise, and not on the basis of whether they integrate with an existing range.

Lower development costs. If all equipment adheres to the same basic international standard, there are no costs incurred in the development of new equipment standards.

Fewer development risks. Because all the equipment developed would comply with a recognised international standard there are no problems in trying to overcome user acceptance of a new product.

Wider choice of development staff. Eventually it is likely that almost all the equipment vendors will be working on the development of X.400 products. This implies that there will be many talented people employed on such projects. In the long term this will give vendors a wider choice of experienced development engineers from which to recruit their staff.

Summary

It should now be clear that the adoption of the X.400 solution can hold the key to a number of benefits for both the user and the supplier. Certainly the suppliers, both of services and products, have been quick to realise the potential of this new standard and evidence of this is given within various chapters of this book. This service provider migration is doubly interesting because of its scope, and it now seems that all the major First World countries around the world are carrying out some form of activity. This points the way, ultimately, towards the establishment of a worldwide messaging infrastructure.

The final element which is required is that of user adoption of X.400. This has been slow up to now but is likely to increase rapidly as the X.400 products and services become readily available. When this has been achieved the true benefits of this approach should become really apparent.

8 Implementing X.400 Within Your Organisation

THE PLANNING STAGE

The implementation of X.400 within an organisation is a complex task which requires much 'ground work' before getting to the stage of a working installation or even a pilot scheme. The essential approach with X.400 implementation, as with any other form of Office Automation (OA), is to integrate it with the organisation's total OA strategy. Only when any form of OA is fully integrated does it achieve its full potential with respect to cost and efficiency performance.

This planning stage is particularly crucial with X.400 as it can act as a means of integrating/interconnecting other technologies and is a flexible basis for many applications in addition to just electronic mail. In this way message handling needs to be considered in the widest possible context so as to establish all the possible ways in which it can be used. This activity in itself will help to justify the cost of such an installation.

The flexibility which has been built into the message handling recommendations means that there are a number of ways in which any given communication problem can be solved using X.400. Because of this, careful consideration of the type of installation is required and in particular its external connections, such as to other private messaging domains, a public service or both. The installation itself is, however, generally constrained by existing equipment investments and the need to preserve these.

X.400 message handling is the first readily available user application to be based upon the principles of Open Systems Interconnection (OSI – see Appendix 3). The OSI architecture is a complex seven-layer structure within which each of the layers carries out a specific function with the use of an appropriate protocol. OSI standards have been designed to be as flexible as possible and to this end contain a large number of options of which only a certain percentage are required to achieve an acceptable interworking solution. This standardised approach in itself presents problems with implementations which have been sourced from a number of manufacturers. Implementations tend to differ either because of ambiguities in the base standards or differences in the interpretation of them. Functional standards (see Appendix 5) are an attempt to produce working/compatible implementations by taking an effective slice through the options and facilities. When procuring X.400 implementations care has to be taken to specify the appropriate profile for a specific application.

The general move towards electronic message-based communications, which is advocated in this book, can sometimes lead organisations into difficulties. The legal status of electronic messages is not yet clear and hence this also needs consideration during the movement away from postal communications which is implied by implementing a complete electronic messaging system.

All of the preceding factors need careful consideration when looking at the potential for X.400 message-based communications within an organisation. It is only by such a stringent approach that the real benefits of this technology will become fully apparent.

WHAT ARE THE OPTIONS WHEN IMPLEMENTING X.400?

The architecture chosen in the development of the X.400 Message Handling System (MHS) model, ie discrete functional blocks, allows plenty of flexibility when carrying out implementation. The two main functional entities of X.400, the User Agents (UAs) and Message Transfer Agents (MTAs), can be co-located or separated, so there is scope for their implementation to be on the same or separate hardware. This general flexibility allows X.400 concepts to be integrated with existing implementations so that they can provide a homogeneous interface between them.

In order to consider clearly the choices for an organisation wishing to implement X.400, all the potential options must be viewed at a general level. When this has been achieved, specific solutions for the problem defined can be selected based on its practicality and cost effectiveness. These options can be subdivided quite neatly into three distinct headings:

Possible interactions. The possible interactions between X.400 management domains.

The means of interconnection.   What, in general terms, are the ways to realise X.400 connections?

What hardware and software?   What types of existing hardware can be utilised with X.400 and what are the implications with X.400 replacement of or co-existence with current electronic mail software?

These headings are now considered in more detail.

Possible Interactions

Within the MHS recommendation only two types of interaction between management domains were defined:

— Private Management Domain (PRMD) to Administration Management Domain (ADMD) (and vice versa);

— and ADMD to ADMD.

Because of the service spectrum covered by the CCITT, ie administrations and service providers, they did not consider the potential for direct inter-PRMD links. This gap was filled by the ISO MOTIS work which extended message handling features into the private domain. The ISO MOTIS work is important in the context of implementation as it forms the basis of the major European CEPT and CEN/CENELEC functional standards for message handling. So in reality the interactions which can be considered in the messaging environment are:

— PRMD to PRMD;

— PRMD to ADMD and ADMD to PRMD;

— ADMD to ADMD.

In terms of user connections the first two are of most interest, while the third has a more indirect bearing. The first consideration for the user is whether to carry out messaging with another private message system, subscribe to a message handling service, or use a combination of both. The deciding factors here will centre around the following considerations:

— average volume of messages on a daily basis and whether these are to many destinations, a small number or one in particular (eg linking two branches within a company);

— are the added value features (eg telex/telematic access, international messaging connections, etc) of a public MHS service, such as Gold 400, of significant value to an organisation?

If an organisation can foresee that a high volume of messages will pass between their own site and a small number of others the charges of a message handling service may be excessive. Instead, it could find that it is more cost effective to link directly using X.400 protocols over either public or private data networks, or possibly a combination of both. X.400 Message Handling Systems are based upon X.25 packet switching protocols, at their lower layer (see Appendix 7) therefore it is possible to use either British Telecom's Packet Switch Stream service or a private network formed from proprietary X.25 devices.

Looking at the other side of the coin, an organisation might currently receive electronic messages via a number of existing messaging media. In such a case it could be attractive to streamline their communications with a single connection to a public message handling platform which would allow access to most forms of messaging media.

The requirement for efficient international communications is of paramount importance to some groups. In this context a connection to a public message handling service will certainly offer more potential for international communications than with any other service.

In terms of a direct cost comparison between connection via a public MHS service and direct connection via public data networks the UK market provides probably one of the first clear indications. The charges for BT's Gold 400 message handling service and for PSS are indicated in the tables below. It must be pointed out, however, that access to Gold 400 is itself via PSS so the charges for connection to this service must be considered as additional to those charges incurred operating a direct link. This obviously means that the Gold 400 solution is more expensive: however, the potential customer has to weigh this extra charge against the facilities which connection to the service will yield.

Access to Gold 400 is via a Dataline 2400 or 9600 PSS connection, for which the charges are as follows:

— Dataline 2400, giving a line speed of 2400 bps full duplex;

— Dataline 9600, giving a line speed of 9600 bps full duplex.

The charges for these circuits are:

Dataline	Connection charge	Quarterly rental
2400	£850	£525
9600	£950	£925

In addition to these will be the charges made for the calls via the PSS to Gold 400:

Datacalls – Inland Call Charges					
Volume charge per kilosegment			Duration charge per hour		
Standard	Cheap	Low	Standard	Cheap	Low
£0.30	£0.20	£0.185	£0.30	£0.20	£0.185

All the charges detailed so far apply when considering a Gold 400 connection, therefore care should be taken if a clear idea of the total cost is to be gained. The most useful approach is to consider the type or volume of messages to be transferred on a daily basis, work out the average cost, and then contrast this figure with the benefits of an X.400 ADMD connection. Certainly the charges for the public MHS service are not cheap, but in the long term it may offer some desirable facilities.

If only direct connection between PRMDs is required then only the PSS charges will be incurred, or those of some other form of direct data circuit connection. The comparative costs and/or benefits of each solution would have to be analysed on the basis of the applications of individual organisations in order to consider which approach is most suitable. A final point which should be mentioned is that a private network of X.400 connections is inherently more secure by nature, ie no third party intervention in the process.

It should be mentioned that with both Gold 400 and direct connection scenarios, the PSS line can be used for other data applications in addition to X.400. Hence, it is up to the individual organisation to make the best use of the PSS facilities.

The Means of Interconnection

Once the type/s of interactions required to meet an organisation's needs have been established it is possible to consider how the X.400 connection will be implemented. In general, organisations considering migration to X.400 message handling will probably already be operating some internal form of electronic messaging. In such a situation the options fall into one of two categories:

— the provision of an X.400 package will be the basis for internal and external messaging needs, ie a new package which will reside on existing hardware;

— the use of an X.400 gateway – these may either be full or half gateways.

An accepted means of interworking between two incompatible networks or systems is the provision of a gateway facility to bridge between them. Such a gateway acts as a user of both networks and provides conversion facilities between the two sets of incompatible protocols. This will probably be the most popular way of providing X.400 facilities on messaging systems which cannot support X.400 in their native mode.

Two types of gateway can be considered, a full gateway and a half gateway. The former implements each layer of the two network architectures being linked and provides a mapping between the protocols being used. It allows interworking between the two systems which is transparent to the users who are unaware of which system is being accessed. Both systems are treated as peers by the gateway and the only restrictions arise when there is no true match between the two sets of protocols.

A half gateway treats the two networks in a different way, with one being looked upon as the primary network and the other as the secondary network. On the primary side full X.400 OSI protocols would be supported, while on the secondary side the system would need to support only the higher level X.400 protocols using the most appropriate lower layer protocols. With this type of gateway there are no limitations imposed on the higher level primary protocols, X.400, would operate end-to-end across the gateway. The gateway facility merely has to map the lower layer protocols of each network.

Those are the general options when considering connection to the X.400 environment from an existing messaging system, but it may be that an organisation feels that it will be more cost effective to completely replace an existing system. In such a case the obvious solution is a new 'turn-key' system which offers X.400 compatibility.

What Hardware and Software?

It was stated earlier that in most cases organisations considering migration towards the use of X.400 protocols will probably have existing hardware and software investments which constitute their current means of electronic communications. This situation obviously makes it desirable to preserve as much of the existing investment as possible when migrating to X.400.

Recent years have seen a rapid increase in the types of computing hardware available for the office environment within organisations. Examples of such equipment would be:

Mainframe computers. Usually provided on a corporate basis to cope with all needs; other systems connected could be minicomputer/s, local area networks (LANs), standalone microcomputers and dumb terminals.

Minicomputers. Smaller departmental systems which run a group of either dumb terminals or microcomputers, generally the basis for an integrated office system facility.

Local areas networks (LANs). Small high-speed networks, provided for a reasonable size of office, with a dedicated file server. Connected to these networks would be microcomputers with the appropriate communications card and software.

Microcomputers. The vehicle for the distribution of processing power down to the individual's desk.

All of these pieces of hardware can now be, and in many cases are, equipped with electronic mail facilities or – in the case of the micro – the means to achieve access to such facilities. In this context, then, all of the above can potentially have access to the X.400 messaging community.

It is now possible to purchase portable, ie across many systems, X.400 packages from a number of vendors. These will act as the native-mode electronic mail system for an organisation as well as allowing direct external communication with other X.400 compatible systems, be they public or private. Such packages are available for a complete range of hardware, from mainframes down to micros and covering most in between, and for a number of operating systems. This approach maintains existing hardware investments but will obviously present the users with a different interface to their electronic mail facilities because the existing package has been replaced.

Integrated office systems which combine and integrate many office utilities, such as word processing, electronic mail, diary, etc, have become popular in recent years. The reason for their popularity is that they have yielded access to all the required office functions while packaging them in a nice 'user-friendly' manner. Because of this people who use such systems have become well versed in their operation and their user interface. In such situations it is not desirable – and could even be counter-productive – to completely replace such a package because of the arrival of a new technology such as X.400. The most productive approach is to integrate the new features with those of the existing systems, thereby preserving the existing user interface and merely adding another feature. The vendors of such systems have been quick to realise that this is the right approach and have busily been announcing or developing either new releases of existing software or integrated gateway facilities for existing packages. In this way X.400 facilities are now available with most of the major integrated office systems and those not yet currently provided will be quick to follow.

Local area networks and their associated mail packages will not be excluded from the message handling environment; there are already some X.400 packages and gateways available for the file servers. In the case of a complete X.400 package, this would allow the file server to assume the role of an MTA while the PCs would require additional software to perform the duties of a UA.

Now that the general options for X.400 implementation have been discussed it is time to think about the stages and implications to be considered during a practical implementation.

THE STRATEGIC APPROACH TO IMPLEMENTATION

The danger of proceeding with the implementation of OA without some kind of strategic approach cannot be over-emphasised. Short-term benefit should not be the sole criterion for new systems development; rather the long-term view should be taken in terms of the ultimate progression of the system as a whole.

Within organisations a strategy is sometimes required to overcome the internal inertia and resistance to change. Without this approach, little pockets of new technology will spring up here and there. There will be no co-ordination and the full integration of the various technologies will not be achieved.

Building the crucial communications infrastructure within an organisation will also be addressed by the strategy. This will provide the interconnection medium between the various systems within the office, such as workstations, computers, word processors, phototypsetting machines, intelligent photocopiers, viewdata, telephone sets, PABXs and many others. This aspect of the strategy is particularly pertinent to X.400 and will be considered in more detail later.

Another point to be considered is the need to ensure that appropriate resystemisation of office procedures, including staff reduction (if necessary) and redeployment programmes, takes place. The training of staff is also crucially important if the technologies are to be fully exploited.

The introduction of new technology has to take account of the organisation's corporate or business plans, existing computer systems, systems under development and systems planned for the future. In particular, it is necessary to ensure that short-term activities and investment are compatible with long-term needs.

Finally, a strategy is needed as a vehicle for senior management backing and commitment to new technology, which is so essential for its successful introduction and on-going usage by employees at all levels. In some organisations, of course, the thrust for introducing new technology will come from senior management themselves as they will have formed the view that it is in the organisation's best interests.

Including X.400 Within a Strategy

When it has been decided that X.400 message handling should have a role to play within an organisation it requires a reassessment of their current OA strategy or, if none exists, the development of one. The scope of this technology is so broad that it requires a co-ordinated approach with substantial commitment if its introduction is to be successful. X.400 protocols can form the basis of a complete interconnection/integration network within an organisation. Further, as worldwide X.400 service interconnection/integration develops, it is likely that adoption of this one technology could fulfil all a company's external communication requirements.

Consideration has to be taken of all existing means of intra/inter-organisational communication in order to assess where the introduction of message handling services is likely to have the greatest impact. This approach takes into account that the introduction of such a technology is likely to be on a gradual basis rather than trying to adopt some form of blanket technique. In all cases a clear application should be identified. It is not sufficient to merely 'want' new technology; there has to be a definite use for it. In this direction the scope of X.400 is almost without limit and it is only when people look at their own requirements that new applications will become apparent.

Because message handling is based on the concepts of Open Systems Interconnection (OSI) there are also wider issues worthy of consideration. The concepts of OSI can support other applications, such as file transfer, virtual terminal, etc (see Appendix 4). These should also be taken into consideration, because although most are not yet as mature as the message handling application their usage will become commonplace in the future. Exploitation of the OSI framework is obviously desirable in achieving the cost-effective use of this technology. With this in mind it might affect some of the product purchases for X.400 by using a supplier who will be able to also supply other OSI applications when they arrive.

PILOT SCHEMES

Small pilot schemes of OA are a relatively safe way of gaining experience with new technology – measuring its potential impact, productivity gains and other related factors. The pilot scheme would also act as a feedback loop to the strategy in order to modify it in the light of practical experience. There are basically three types of pilot:

— short-term experimental pilot projects which are discontinued after experience has been gained;

— pre-production pilots which provide a small-scale demonstration of the final system;

— hybrid pilots which can combine features from the last two project types.

Most pilots carried out recently fall into the hybrid category, which have generally combined a common set of features:

— project conducted in a closely controlled environment in order to allow careful evaluation of results;

— the pilot will influence the final design in terms of refinements;

— a significant degree of measurement and evaluation is required;

— success is hoped for but not assumed in advance.

Pilots can be of great use where it is difficult to establish the justification for a project in terms of cost and performance, without some practical experience. Indeed, some organisations consider that a pilot is the only means through which OA can be successfully introduced.

X.400 is a new technology and it may well be viewed with some suspicion from departments within an organisation. In such cases the use of a pilot scheme may be essential in gaining approval from certain quarters, particularly management who have to approve the expenditure for such schemes. With an X.400 pilot it may be relevant to include the people who doubt its potential in order that they may be exposed to the technology on a 'first-hand' basis. The use of such a well-devised pilot will be instrumental in overcoming the general inertia from certain quarters to the introduction of new technology. Finally, the pilot will also be a useful test of your chosen suppliers and will give an indication of whether or not they will be able to meet any of your future X.400 requirements.

SPECIAL CONSIDERATIONS WHEN PROCURING X.400 PRODUCTS

X.400 message handling systems are based on the principles of the ISO Open Systems Interconnection (OSI) seven layered architectural structure. This structure is intended to act as a basis for many future applications, some of which have not yet been devised. To this end it contains a wide range of options which impart sufficient flexibility in order to achieve this aim. The problem with this flexibility is that it provides the implementor with choices and making the wrong ones can produce implementations which are incompatible. Because the standards themselves are produced in normal language format they are liable to have ambiguities within them which will lead to different interpretations by different implementors.

X.400 systems are not meant to stand on their own; rather their true usefulness can only be fully achieved when they have been connected to other conformant systems. These other systems can be sourced from a number of suppliers and therefore the need for a common understanding and interpretation of the base standards is crucial. To this end the development of standards profiles (see Appendix 5) have gone some way to helping the situation. The profile takes an effective slice through the seven OSI layers, selecting options and placing realistic limits in order to achieve a subset which will act as a basis for working implementations. This approach removes many of the permissible choices but can still do nothing to alleviate the problems of individual interpretation.

The next stage which is required to guarantee interworking between products is conformance and interoperability testing (see Appendix 6). These tests will show the conformance of systems to a base standard, or profile thereof, and their ability to interwork with other implementations. At the time of writing such facilities are not yet widely available, although NCC, BT and the Networking Centre in the UK are offering some services which will be progressively developed. Early to mid-1989 would seem to be the most likely time scale for such services to become widely available. When this occurs vendors will be able to have their systems tested and build up logs of other implementations they are compatible with: this information could then be made available to prospective customers. It is only at this stage that potential purchasers can feel secure in the knowledge that the systems they purchase are able to interwork with others.

For the moment, however, an organisation procuring such equipment is in a very vulnerable position. A great deal of thought must precede the production of a document which details the requirements from suppliers. Always specify the problem in as precise a way as possible, including any reference to future development of the system. Generally customers, when talking to suppliers, leave so many things unsaid, taking for granted what they really should be questioning. Do not fall into this trap because suppliers will usually only answer the questions you put to them. Suppliers have a notoriously bad record in only supplying the information they wish to give: in this way they avoid highlighting any weak areas of their product. It is up to the customer to be forceful and not accept weak and incomplete answers. Question and keep questioning until sufficient clarification has been given. Always be very specific about the use of a system (ie its applications), its interactions with other systems and potential interconnections. This is the key to a successful implementation.

Some interworking tests of X.400 equipment have already been performed on a public basis. Examples here would be:

— the ceBIT 87 X.400 demonstration;

— British Telecom's engineering trials of Gold 400;

— Telecom 87.

These have all provided the vendors with useful knowledge about the interconnection of their respective products, so it is not exactly a 'green field'. Hopefully they will be able to draw on this experience to provide early customers with an appropriate base for future development and interconnection.

Reverting to the question of profiles of standards, and in particular X.400 ones, it is crucial to specify the appropriate one for your requirements. There are three major 1984 X.400 MHS profiles which are being referenced by vendors at present:

CEPT A/311 – This profile carries the European Pre-Norm. (ENV) number of ENV 41202. It specifies the functions and procedures for the interworking of private (PRMD) and administration (ADMD) domains. Because of its ENV status it is the major European standardised profile for PRMD ADMD interworking.

CEN/CENELEC A/3211 – This profile also holds ENV status but covers interworking between PRMDs only. ENV 41201, as it is known, is the major European standardised message handling profile for PRMD to PRMD connection.

NIST NBSIR 86-3386-5 – This is a North American profile for X.400 message handling interworking. It specifies the functions and procedures for PRMD to PRMD and PRMD to ADMD connections.

It should be fairly obvious, from the above, that for European X.400 systems either ENV 41202 or ENV 41201 should be specified, depending on the requirements of a particular installation. A recent interesting development in the profile work has been initiated by the UK government. The Government OSI Profile, or GOSIP, attempts to provide a procurement specification for all UK government OSI systems. GOSIP, which encompasses MHS, is likely to be very significant within the UK.

When dealing with the procurement of any OSI-based equipment the points mentioned here cannot be over-emphasised. In the long term they may be crucial when attempting to develop a system further so as to allow greater interconnection/integration and additional applications.

THE COMPLETE PLANNING, IMPLEMENTATION AND DEVELOPMENT CYCLE

So far, the major elements in the planning/implementation of an X.400 system, the strategy, a pilot, and procurement, have been considered, but what is the overall picture and how do these relate?

The first point to consider before any work can be carried out is who will sanction it. The realisation of the need for a coherent approach to implementation may indeed originate from managerial levels in an attempt to ensure cost effectiveness. It is also likely that someone at a lower level may have a better feel for the impact of such systems and believe that a strategic approach is more beneficial. In either case a high-level decision is required in order for work to commence.

Figure 8.1 indicates a simple representation of an office automation (OA) planning, implementation and development cycle which is applicable to any specific case such as X.400. From this it can be seen that the first stage of any project has to be an identification of the areas of opportunity where developments or improvements can be made. This can basically be defined as perception of an application for technology. The obvious follow-on step from this is to select the technology which will best achieve the specific objectives which have been defined in the initial stage.

As with any project in business the implementation of OA has to be cost effective, otherwise there would be no incentive for its adoption. Cost/benefit justification can be achieved in a number of ways: it may utilise information from the achievements of industries in a similar sector; input from a pilot study; previous OA experience. All of these will form valid input to justification of the approach taken in the strategy.

So those are the main areas of input to the strategy. But how will the strategy itself be formulated? There are a number of factors to be considered here, such as:

— the stages of implementation;

— supplier selection criterion;

— how each stage will be evaluated for success or otherwise and how this information will be utilised as input to the rest of the project;

— requirement for a pilot scheme and how knowledge gained affects the evolution of the strategy;

— who will have to judge the strategy on whether it has sufficient merit to be pursued.

All of the above points and more should be considered within the strategy document. (For further detailed information refer to NCC publication *Planning Office Automation – Electronic Messaging Systems,* by J A T Pritchard and P A Wilson.)

After the strategy has been produced, the next stage may be a pilot scheme. If this is so, it will require the selection of a supplier which, in the case of an X.400 implementation, should be based on the guidelines detailed under the heading 'Pilot Schemes' (p. 128). The choice of suppliers may be changed in the light of experience at this pilot stage of the implementation; it will depend on their performance.

When the final choice of supplier/s for the project has been made the implementation itself can be carried out. When completed, and after a period of acclimatisation, some kind of audit should be done to assess whether the expected gains have been achieved. If the implementation has not been as successful as anticipated it is possible at this stage to feed back some experience into the strategy in order to influence the later stages of implementation.

THE LEGAL IMPLICATIONS OF ELECTRONIC MESSAGING

One aspect in the introduction of electronic messaging which is often overlooked is its legal implications. Electronic messaging, when successfully implemented, should become an integral part of all aspects of

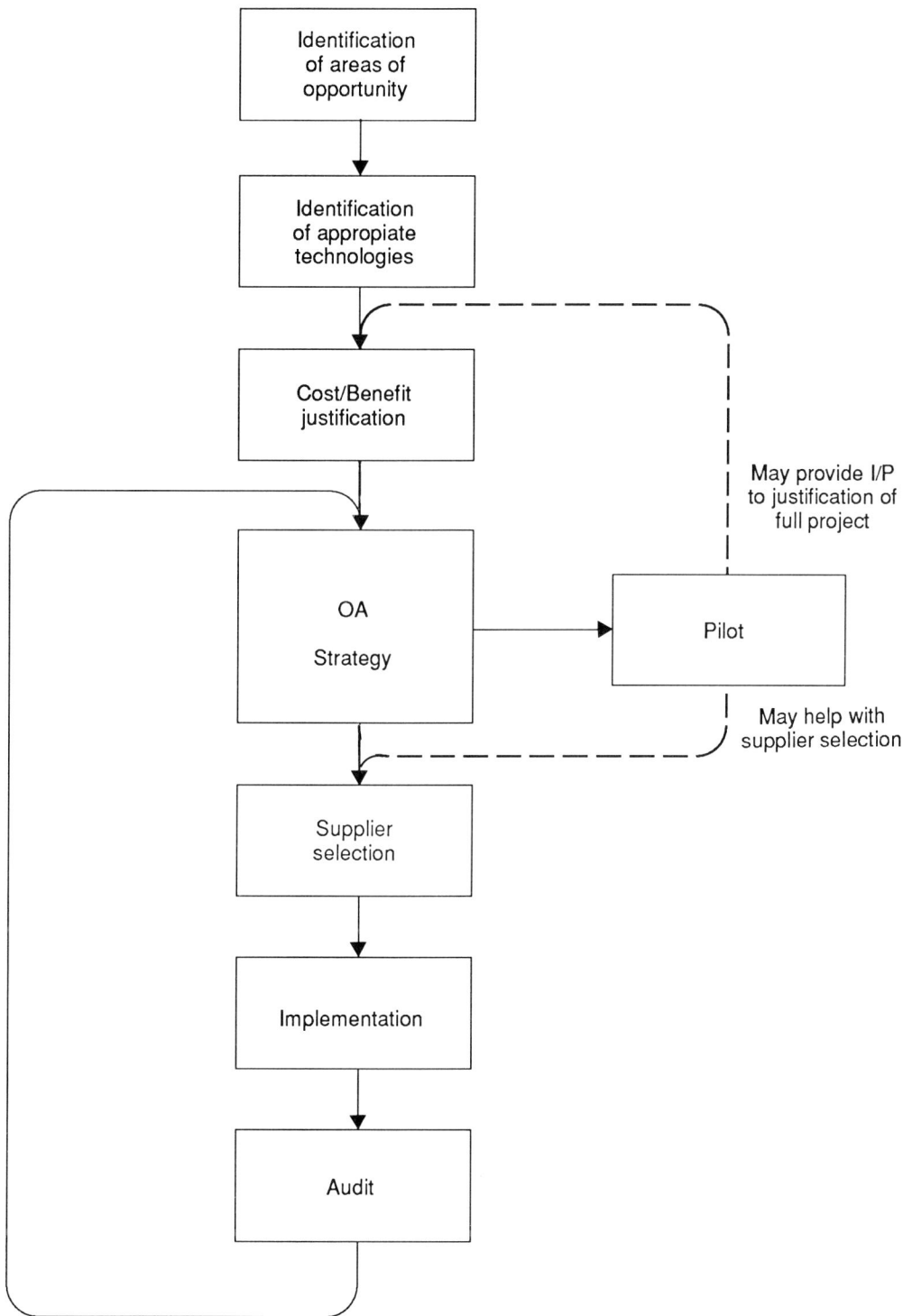

Figure 8.1   The OA Planning, Implementation and Development Cycle

business operation. One, often quoted, application for electronic messaging is its use in the formulation of contracts between organisations. Electronic messaging is an entirely suitable media for this application but can raise legal issues when problems occur. At present, neither legislature nor the courts have addressed this area and hence it is up to organisations operating trade links to establish their own guidelines. Such issues should be considered as early as possible in the implementation cycle so as to smooth the introduction of this technology.

Within contract law, times and places can have legal significance and it is just such details that are difficult to establish clearly when using the store-and-forward concepts of X.400. The time when a message was passed to a system for transfer has no fixed relationship with the delivery time, hence unpredictable delay is introduced. The delivery of a message and receipt of it by the user also have no fixed relationship. A message delivered to the recipient's system may remain there for many hours or even days before the recipient himself actually logs on to the system to be able to read it. In the intervening period the conditions of the original contract may have changed. This can lead to a situation where the recipient decides to take up the conditions of the original contract at a stage when the offeror believes he has withdrawn them.

Until either a legal framework or commercial practice establishes an appropriate set of guidelines it falls upon organisations, the users of electronic messaging, to establish sensible rules. This could be quite easily achieved with advice from a legal department and would involve analysing the precise mode of operation when contracts are to be established. These guidelines would state exactly the information to be contained within a contract and the required answers to be expected. It would also seem advantageous to make these guidelines available to trading partners in order to inform them of the practices which are to be adopted.

In summary, then, the legal aspects of electronic messaging should not be overlooked when moving towards its widespread usage. Failure to address this issue could result in problems at a later stage which may be difficult to resolve.

Conclusion

It has been adequately stated within this chapter that X.400 implementation has to be approached with care. Identification of an application and the means with which this can be achieved are just the first step. The selection of the right supplier for the job is undoubtedly the most difficult stage. Accurate specification of the solution required, its potential applications and future development is necessary to ensure a minimum of wasted effort. Successful completion of this activity is the only way to ensure that the implementation itself meets the original criteria laid down.

The aim of this chapter has not been to dissuade organisations from implementing X.400; far from it. Rather it should have highlighted sufficient problem areas for it to be approached in a sensible manner. It is by virtue of this care being exercised at an early stage that will, ultimately, allow the full benefits and potential of X.400 to be appreciated.

9 X.400 Implementation Case Studies

The previous chapter detailed some guidelines which it would seem sensible to adopt when approaching the complex topic of X.400 implementation. With these in mind this chapter contains two case studies of message handling implementation. The first of these follows the process through from start to finish (ie specification to implementation); the second, which has reach the stage of a detailed migration strategy, is currently being implemented.

The most interesting aspect of both these case studies is that X.400 seems to have found a common application. Each organisation is using message handling systems not only as a means of external communication but also as a basis for the integration of existing internal systems so that they form a cohesive whole. This, at least initially, is likely to be one of the major applications for X.400 systems. The two case studies detail:

— the use of X.400 as a means of linking the messaging systems of UK Civil Service departments;

— the use of X.400 as a means of integrating the existing forms of electronic messaging within a large European private company.

CASE STUDY 1:

INTER-DEPARTMENTAL ELECTRONIC MAIL USING OSI AND X.400

(*Author's Note:* This case study was produced by William H McKinley who, at the time of writing it, was the IDEM Development Manager for the CCTA. Many thanks to Bill for his efforts.)

Summary

In May 1986 the Central Computer and Telecommunications Agency (CCTA) was awarded the contract for the pilot Inter-Departmental Electronic Mail (IDEM) System. This system is now the hub of an electronic mail network to serve the UK Civil Service. It uses OSI and X.400 protocols to allow the transfer of mail between the central IDEM System and user systems in government departments. IDEM was the first government procurement in which OSI, including X.400, has been specified as a mandatory requirement.

The CCTA Initiative

In early 1984 the CCTA commissioned a feasibility study on the requirements and potential for inter-departmental electronic mail in government. The aims of this study were to specify the requirement for non-voice electronic communications, and to set out a range of options for meeting it. The study was genuinely inter-departmental, with a CCTA/Logica study team working under the directions of a Steering and a Consultative Group each comprising senior civil servants representing all the major departments – the potential users.

The feasibility study reported in September 1984 and the findings were published by HMSO as No.9 in the series *IT in the Civil Service*. In summary, the study revealed a wide interest in, and demand for Inter-Departmental Electronic Mail (IDEM), and identified a large number of inter-departmental links and 'communities of interest' who could be early users of an IDEM service. It was also found that the demand for such a service was beginning to be met in the form of ad hoc bilateral experiments. There were also widely expressed needs to avoid 'lock in' to any manufacturer specific protocol, to avoid the installation of incompatible systems and to provide a solution which yielded a strategic migration path to a government-wide electronic mail system predicated on OSI.

Looking at the wider expenditure by government on IT products and services, the study suggested that some form of central electronic mail service could increase significantly the benefits of departments' investments in office technology. If the predicted investment is realised, the government will spend some £200 - £400 million on office technology by 1990. It was calculated that a marginal additional investment of, say, £1 - £5 million in electronic mail could bring additional benefits in increased consultation, personal efficiency and improved decision making, as well as cost savings in a number of areas. The study also recognised concern expressed about the need for security features appropriate to the normal office environment.

Design Options

In the feasibility study, a number of options for the IDEM service were drawn up. Some factors were central to the design of all these options:

— interworking with other office systems;

— ease of use;

— appropriate security;

— a requirement for a basic final form of document;

— a requirement for a revisable form of document;

— acceptable cost.

Following the feasibility study, in November 1984 it was decided to proceed with the design of a pilot system. The IDEM service was conceived as a general purpose service carrying material classified as Restricted, Private (eg Commercial-in-Confidence) and Unclassified. The service would be required to support both a dial-in facility for users with standalone terminals as well as links to office systems in departments.

Detailed discussions were held with a range of suppliers. Because of the apparent immaturity of the higher OSI layers in late 1984, early design ideas for IDEM were based on a central system supporting several industry protocols for electronic mail and document formats, these 'standards' being dictated by the systems that would wish to connect. The perceived advantages of such an approach were: the ability to connect each departmental system in its native mode, the concentration of all the risk and development within a single site at the centre, the use of known protocols, and a relatively short development period.

However, some disadvantages also became apparent. Obviously a major function and cost of the centre would be protocol conversion, perhaps with no obvious or easy upgrade path to OSI at a later stage. Nevertheless such an approach appeared feasible and the major decisions seemed to be on the selection of the protocols and document formats to support the users, and, vitally, to design a suitable addressing structure.

X.400 for IDEM

However, as the discussions with suppliers progressed it was found that a large number of them had plans to produce working systems based on the CCITT X.400 Series of Recommendations in the time frame late 1986 to early 1987. The advantages of an X.400 solution became increasingly evident:

— users would be offered a workable OSI E-mail system covering all seven layers;

— IDEM would create an OSI focus, providing a specification for user departments to incorporate into operational requirements;

— as the single largest purchaser of IT products and services in the UK, the government, through the

CCTA, would be in a unique position to make a practical contribution to the development of OSI products.

Putting the onus on the user system to provide an X.400 interface introduced some risk into the X.400 solution, but user departments were keen to see OSI standards implemented and said that they would welcome the opportunity afforded by IDEM to put pressure on their suppliers.

The Central IDEM System

In October 1985 the procurement of the central IDEM system began with the issue of the Operational Requirement. The mandatory features included:

— X.400 service requirements detailing required service elements and user facilities for both Message Transfer and Interpersonal Messages;

— X.400 interfacing protocols, the X.411 P1 and the X.420 P2 protocols, supporting all of the required Interpersonal Messaging Services (IPMS);

— use of OSI; support of the X.411 P1 protocol according to X.410. This specifies the use of:

- X.410 Reliable Transfer Service

- Session service; Basic Activity Subset

- Transport Class 0.

Subsequent discussions about functional standards and 'profiles' with suppliers led to adoption of the May 1986 CEN/CENELEC profile as the basis for the IDEM contract. Because of the special considerations within government regarding security a small number of extensions to this profile were required. These extensions cover issues such as mandatory and defined use of a new P1 'sensitivity indication' (now being incorporated into X.400 1988), and trusted use of this indication by the central system to prevent routeing of sensitive traffic to unauthorised users. There are plans for the pilot IDEM system to connect with public systems such as the Gold 400 service.

Implementation – in Two Phases

Phase 1

In July 1986 an Unclassified mailbox service, ie for Unclassified traffic only, was offered for standalone terminals such as word processors and personal computers (see Figure 9.1).

Phase 2

In July 1987 the Phase 1 system was upgraded by the addition of:

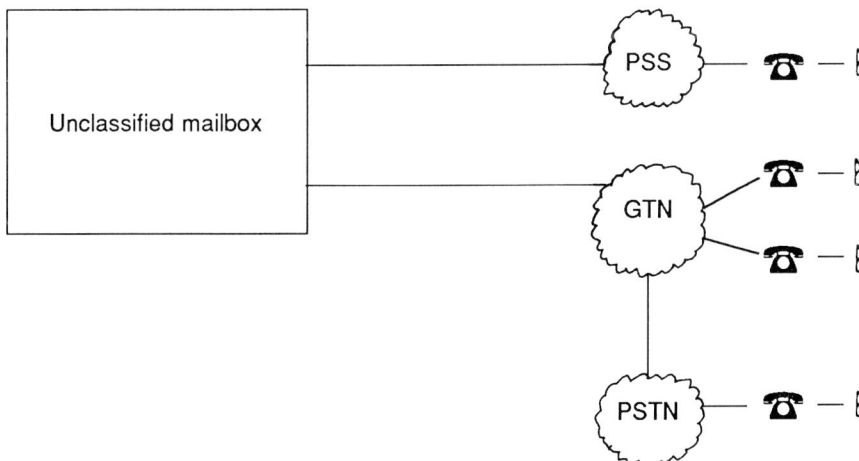

Figure 9.1   IDEM Phase 1 – July 1986

— a Restricted mailbox service to carry traffic that is Unclassified, Private and Restricted;

— an X.400 MHS featuring a Message Transfer Agent (MTA) for the connection of office systems using CCITT X.400 protocols;

— interworking between the mailbox User Agent and the Message Transfer Agent;

— the necessary security features to permit the co-existence of different classifications of traffic.

Phase 2 is shown in Figure 9.2.

Outstanding Issues

Other major discussion points with suppliers during the procurement focused upon naming/addressing, revisable documents and conformance testing.

X.400 draws a clear distinction between naming and addressing. Names are used to describe originators and recipients in terms commonly known about the user (just as in an internal mail system within an office), whereas addresses contain data that directly help the MTA network to locate the user (like a telephone network). It is expected that users will generally use names to route their mail, rather than addresses which will not be so 'friendly'. This implies that there will be some sort of directory or routeing mechanism that will translate names into addresses.

X.400 defines a number of variants for names (see Chapter 3 of this book). The CEN/CENELEC profile states that all User Agents should be at least capable of supporting one of these (X.400 section 3.3.2, form1, variant 1) and defines maximum lengths for each field. It also allows some of the fields to be repeated to cater for a degree of depth in large organisations such as multi-nationals or government departments. However, routeing mechanisms within the MTA network must be able to understand these names.

X.420 specifies a number of different 'body types' that can be inserted into the P2 envelope. The simplest of these is a string of IA5 (ASCII) text, and this is supported in IDEM today. The original IDEM specification also called for support of X.420 Simple Formattable Document (SFD): this requirement was eventually dropped as it became clear that even 'simple' revisable text was going to emerge from ISO's Office Document Architecture (ODA) work rather than from X.400.

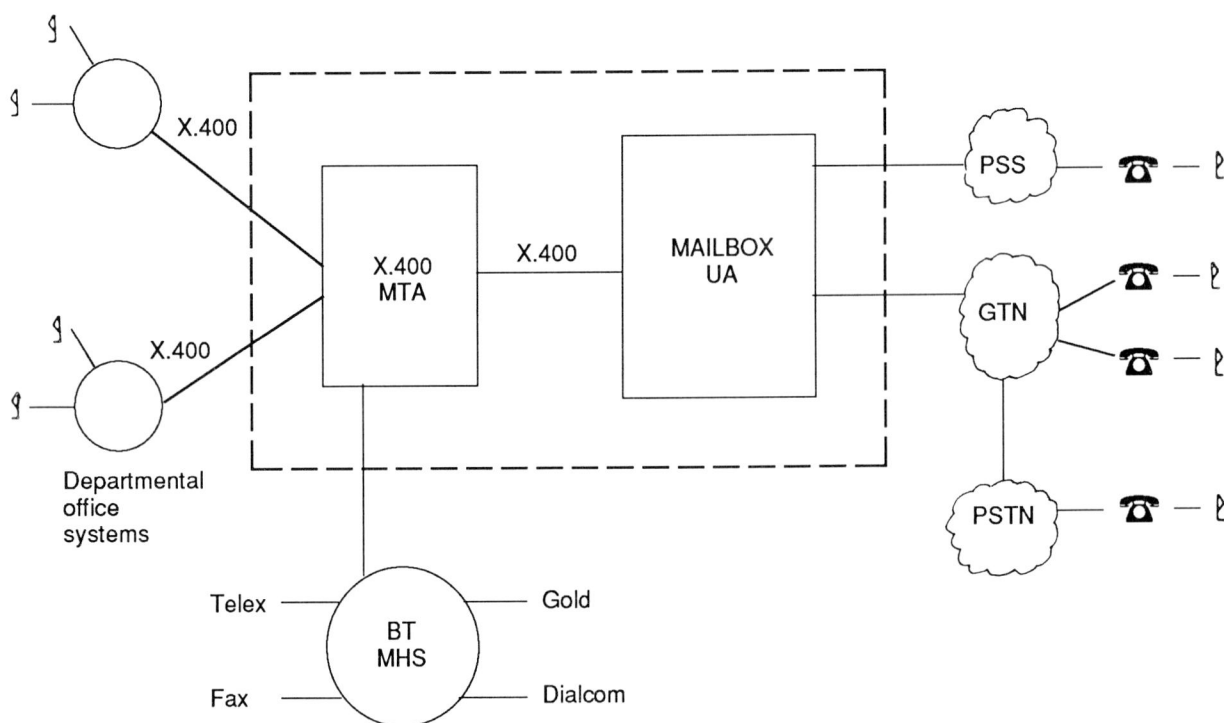

Figure 9.2   IDEM Phase 2 – July 1987

ODA will be the OSI standard for document content, though other document architectures can be sent within X.400 messages between User Agents able to understand them.

The Phase 2 IDEM system entered operation before independent conformance testers were available. Both suppliers and users will welcome independent testers, but these should not be viewed as the complete answer to all interworking difficulties. The X.400 community has benefited greatly from the ceBIT 87 and Telecom 87 interworking demonstrations, through increased awareness of practical interworking issues. For connection of remote systems to IDEM a similar pragmatic approach has been adopted. After connection to IDEM a close observation is maintained until there is a reasonable assurance that all is correct.

The user departments need to obtain X.400 gateways for their proprietary office systems, although some may soon be able to support X.400 in their native mode. This last year has seen the emergence of many X.400 products as well as some interesting announcements.

Maturing IDEM

As the IDEM service grows from a pilot with a single centralised MTA to a network of distributed MTAs, there will be a number of issues to be resolved. Initially these issues will probably concern conformance and integration, at least until conformant versions of X.400 are widely available. There will also be problems with naming and addressing as the connected systems grow, and in turn connect to other systems; because of its position in the hub of the network, the pilot IDEM system is in a very strong position to control this change.

But as the use of X.400 and OSI in government grows, IDEM will need both to decentralise and to connect to departments with large, comprehensive networks of their own, which could quite easily dwarf the original IDEM network. During this growth the CCTA in its central role as IDEM naming authority will need to control the naming/addressing structure of all systems within the IDEM domain, and to co-ordinate the structure of other private domains within government that wish to connect to IDEM and to each other.

Conclusions

Having taken the decision to use OSI and X.400 for IDEM, and following the successful procurement of the central IDEM system, the CCTA is convinced that both OSI and X.400 are achievable now. X.400 offers many benefits to organisations with substantial electronic mail and information technology requirements. There is almost nothing about X.400 that is difficult to understand, and in its basic method of operation it is identical to the conventional mail services with which all organisations are familiar.

With such a good measure of agreement between large users, PPTs and the international standards bodies, the path is now clear for the IT vendors to devote their expertise and resources to bringing the products to market – the user demand is already there.

Footnote

Subsequent to the production of this case study it was announced that the IDEM project was to be ended on the 31st March 1989. In 1986, when the project was initiated, X.400 messaging and electronic mail were relatively new technologies and government departments were uncertain as to their feasibility. The IDEM clearly established that electronic mail, using X.400 standards, is a reality today which is supported by many suppliers. The purchase of departmental systems using the Government OSI Profile (GOSIP) specification for X.400 now ensures direct interworking between systems using an X.25 base. In the light of these factors it was deemed that the IDEM project had fulfilled its original objectives and hence reached a natural conclusion.

The major achievement of this project can be summarised as follows:

— IDEM offered two main services. First, it provided a dial-up mail box service broadly similar to Telecom Gold. This was used by some 250 users in various departments, and fostered the use of electronic mail within and between departments on a relatively modest scale.

— Second and more important, IDEM set out to link departments' office automation systems by encouraging the development of X.400 links at a practical working level. Such links were successfully made to systems in DTI and HM Treasury, and to CCTA's in-house system MITSY. A large number of connections to various types of systems and organisations outside the Civil Service were also tested, helping to establish the credibility of the open systems approach at a practical level.

CASE STUDY 2:

THE HOECHST STRATEGY OF MIGRATION TO OSI

(Author's Note: The following case study was produced from material which was gratefully received from Dr Harald Nottebohm of Hoechst.)

Introduction

The Hoechst group operates worldwide producing a large variety of chemical and pharmaceutical products. Their manufacturing operations are spread over 64 countries and they currently export products to a total of 170 countries. Hoechst sales in 1987 amounted to 37 million D-marks.

In order to support such a vast manufacturing operation it is obvious that Hoechst require a coherent strategy for all communication and information related activities. With respect to information technology their goal is to make an efficient and future-oriented use of all data processing and communication opportunities. They believe that the method of achieving this goal is by strengthening the decentralised DP-usage. In a company of the size of Hoechst with its numerous applications, which have widely differing requirements, systems need to be optimised for specific local needs. Consequently this leads, indeed has already led, to a multi-vendor environment which is spread widely throughout the organisation.

Hoechst soon realised that only by unification would the decentralised approach make sense. This can be summarised in two statements:

— they need to be decentralised on the operational level;

— they need to be centralised on the strategic level.

Hoechst Communications Strategy

These two statements were then applied to the Hoechst communications strategy in order to ascertain some sense of future direction. This resulted in two basic theses:

Thesis 1 — The various existing networks shall be integrated so as to form a universal network which is based on international standards, which means OSI.

Hoechst, however, realised that not all the standards and corresponding products which were needed are presently available. To account for this situation a second thesis was proposed:

Thesis 2 — The process of standardisation and product development has to be accelerated.

The present situation of networks within the Hoechst group is indicated in Table 9.1. It shows that there are almost independent vendor-specific networks which are used in conjunction with various local area networks, public networks and leased lines.

The vendor-specific networks offer a high degree of functionality while their amount of standardisation is low. Some of the vendor-specific products use the lower layers of ISO OSI model (ie X.25 or OSI LANs) while others are almost entirely proprietary. Figure 9.3 attempts to capture this concept graphically.

Hoechst found that systems based on OSI standards are evolving but at present are only offering modest functionality, eg message handling without any other functionality. They believe, however, that these will develop towards higher levels of functionality (eg combinations of file transfer, virtual terminal, etc) until the ultimate target systems (again, see Figure 9.3) of both high levels of standardisation and functionality are achieved.

Hoechst, like many other multi-national user organisations, already have to exist in a heterogeneous system environment. In addition, the use of third party network systems only compounds the problem. Taking these factors into account leads to the conclusion that OSI-based systems are in fact the

Vendor-specific	SNA (Systems Network Architecture, IBM) DNA (Digitals Network Architecture, DEC) DS 3000 (Distributed Systems, HP) Transdata (Siemens)
Public	Datex-P (Packet Switching Network, X.25) Datex-L (Circuit Switching Network, X.21) Leased Lines Circuit Switching Lines
Private Carrier	Mark III Network (General Electric)
Local Area Networks	Baseband, Broadband, Twisted Pair

Table   9.1

only method by which the provision of full functionality communication will be possible in the future. Gateway solutions will not offer the required degree of functionality because the vendors will not be stimulated to enhance their applications to incorporate all common features because no standard definition of these would exist, eg no standard for vendor A to B communications.

Hoechst do not expect that existing systems will evolve along the horizontal line in Figure 9.3, rather OSI migration will lead them through an interim path, or valley of lower functionality. The benefits for them, however, will still be great because they will avoid the increasing costs of internetworking different systems. Internetworking for Hoechst is an absolute necessity for the future if Computer Integrated Manufacture (CIM), Electronic Data Interchange (EDI) and processable document interchange based on

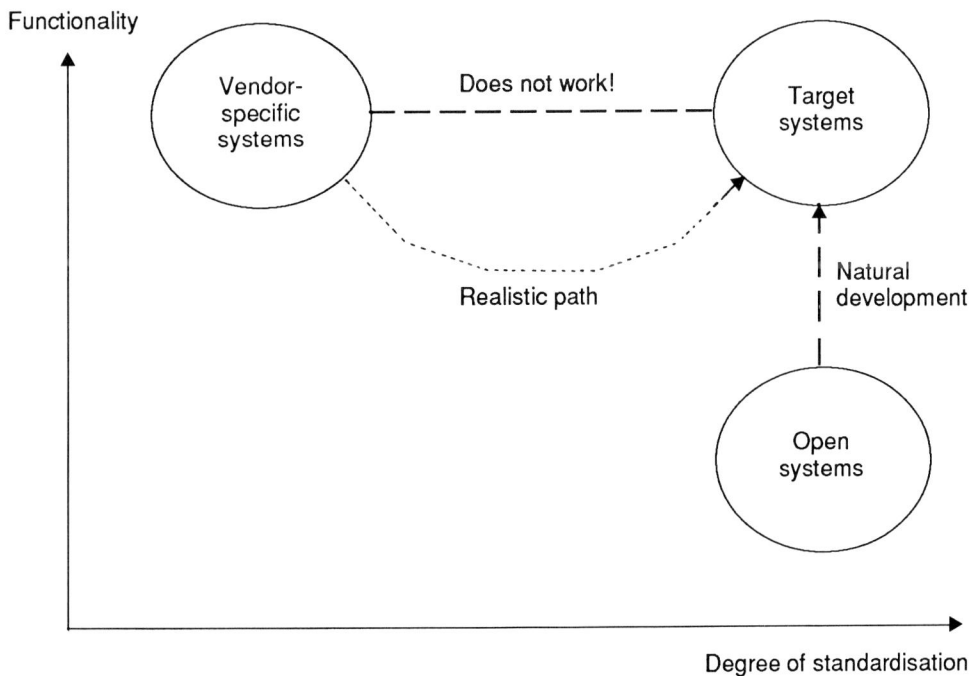

Figure 9.3   Migration to Open Systems

ODA/ODIF (see Chapter 6) are to become a reality. If these can be based on long-lasting international standards Hoechst can effectively protect their immense investments in networks.

Hoechst believe that migration to OSI needs a set of rules to follow:

— wherever possible products conforming to international standards should be used, eg X.25, X.400, etc;

— interim solutions may be chosen if their use can be considered as a step in the right direction, eg standards, even if they cover just part of the desired functions;

— another acceptable interim solution is to use vendor-independent industry standards, eg ODETTE FTP (File Transfer Protocol) and TCP/IP (Transmission Control Protocol/Internet Protocol);

— if it is not possible to adhere to international standards or make use of the permissible interim solutions, then special gateways may be used for a transition period.

Having gained some insight into the general OSI migration policy of Hoechst, it is now relevant to look at the specific example of their migration towards X.400 message handling functionality.

Hoechst X.400 Message Handling Migration

At present, Hoechst use many different electronic messaging facilities within their geographically dispersed organisation (see Table 9.2). The systems were, and partly still are isolated; messages could not be transferred from one system to another. Before assessing how to integrate these diverse systems Hoechst established some general principles which any solution would have to adhere to. These were:

— a user should not have more than one terminal on his desk;

— the user should not be compelled to use more than one user interface;

— the existing hardware, which was installed to fulfil other purposes, should be used and a message handling system should integrate with it;

— the final system should allow any of the users of the different systems to exchange messages automatically.

IBM	MVS: VM	MEMO PROFS
DEC	VMS	All-in-1 VAX Mail
HP	MPE	Desk Manager
GE		QUIK-COMM

Table 9.2   The Main Systems used by Hoechst

For Hoechst it was obvious that the basis for interconnecting their different message handling systems should be the CCITT X.400 recommendations. X.400, being an internationally standardised solution for message handling which already exhibits some maturity and levels of support, fulfilled many of the criteria laid down by the company. X.400 standardises the transfer mechanisms for messages while not attempting to address the actual user interface. Therefore, with an X.400 implementation, users can continue to use the interface with which they are familiar. Many vendors of office systems, all of which contain electronic messaging functionality, have incorporated or announced their intention to incorporate transfer facilities conforming to X.400. Table 9.3 indicates the availability of X.400 products which are applicable to Hoechst.

Integration of Hoechst systems began in late 1986 with the installation of DEC's X.400 Message Router

DEC	X.400 Message Router (MRX)	IV/1986
GE	QUIK-COMM/X.400 QUIK-COMM/All-in1 Connector	1989 I/1988
Nixdorf	Targon-Mail/X.400	III/1987
IBM	PROFS/All-in-1 Connector PROFS/X.400 for DFN PROFS/X.400 DISOSS/X.400	 IV/1987 III/1988 II/1988
German PTT	Telebox/X.400	1989
HP	HP 9000/X.400 HP 3000 DM/X.400 Gateway	IV/1987 II/1988
Verimation	Memo/All-in-1 Connector Memo/X.400	I/1988 1989

Table 9.3   Availability of X.400 Products which are Relevant to Hoechst

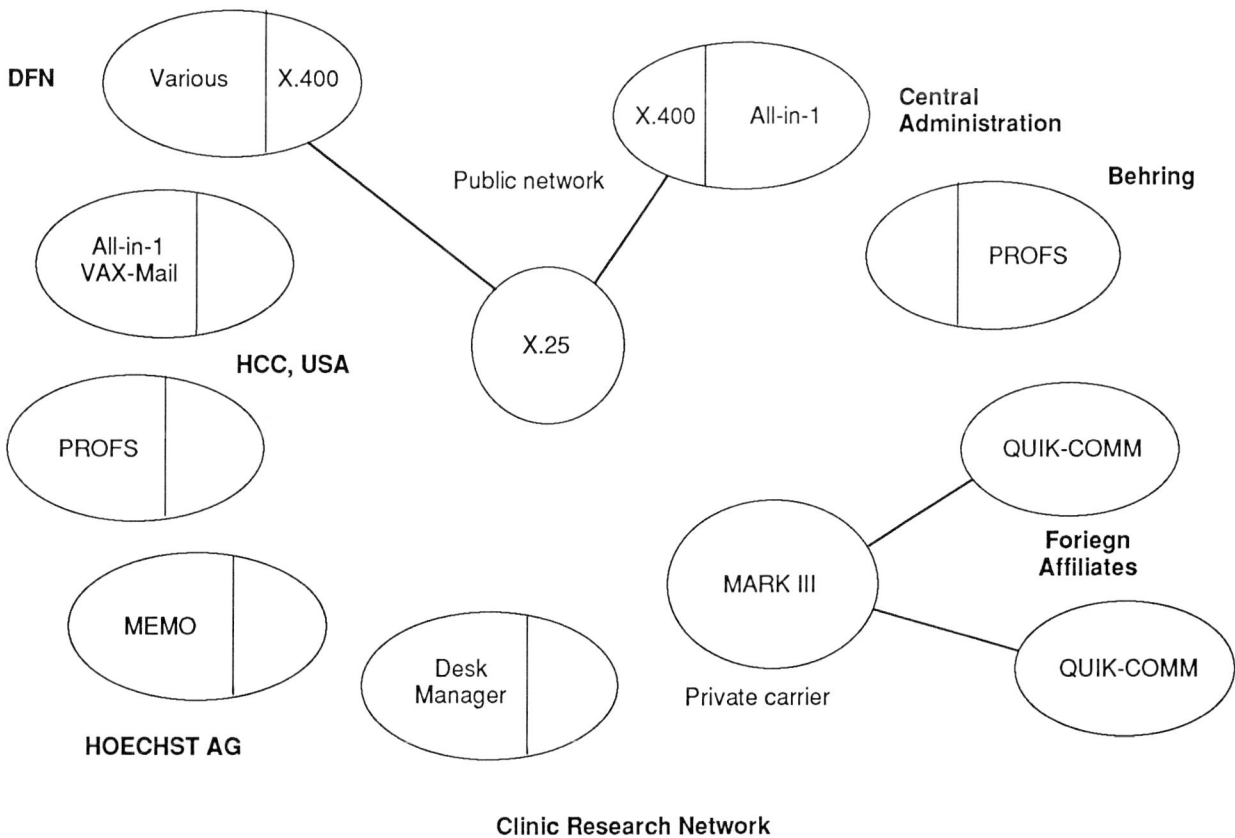

Figure 9.4   First Phase – End 1986

product on the All-in-1 systems of their Central Administration in Frankfurt (see upper right of Figure 9.4). Shortly after, a link was formed to the X.400 German Research Network (DFN = Deutsches Forchungnetz). Via this link it was possible to reach various research networks worldwide.

The other systems in Figure 9.4 serve local purposes, for example, worldwide HP network using Desk Manager as an office system. The GE IS Mark III (see Chapter 5) network is used to move data from smaller foreign Hoechst affiliates. GE IS also provide an electronic mail system called QUIK-COMM used by Hoechst.

In the first phase of the Hoechst X.400 messaging project, there was no way of sending messages from one system to another. The situation at the time of writing has improved substantially, with the addition of three further links, see Figure 9.5.

The US affiliate of Hoechst is the Hoechst Celanese Corporation and they use DEC All-in-1 and IBM PROFS office systems. Since at the time there was no X.400 gateway available for PROFS it was decided to take an interim measure and link PROFS to the All-in-1 system via a 'Softswitch' gateway. The All-in-1 system having X.400 compatibility it could then be linked directly to the Central Administration centre. In Germany, at the Hoechst affiliate Behring, a prototype PROFS/X.400 system, developed by IBM for the DFN, could have been used but it was decided to wait for the full IBM product which was due in late 1988.

The second link which Hoechst established was to the QUIK-COMM electronic mail system by GE IS. Here another software gateway is used, a QUIK-COMM/All-in-1 connector which means that they do not have to wait for the QUIK-COMM/X.400 product.

The third link to Hoechst AG again uses a software gateway, Memo/All-in-1 connector, to allow the Memo electronic mail system to have access to the worldwide Hoechst network.

Summary

This case study gives an insight into the far-sighted communications strategy of Hoechst. Their policy of

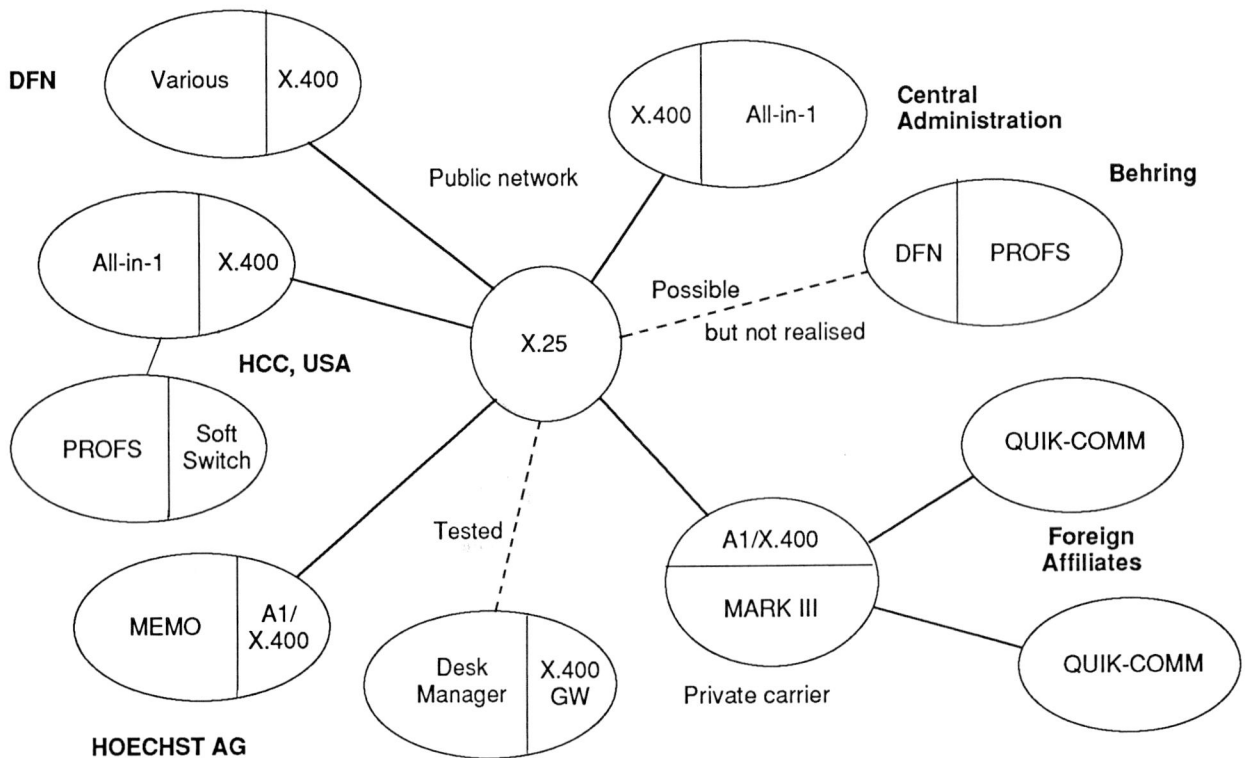

Figure 9.5   Clinic Research Network – Situation mid-1988

gradual migration to systems based on international standards, while realising the lower levels of functionality which are incurred in the short term, indicates that they have truly grasped the issues while not shying away from any short-term problems. Their use of X.400 as a form of 'glue' to bring together their existing diverse systems gives a practical indication of X.400 in action. Apart from merely electronic mail, Hoechst have also realised the potential of X.400 as a base for many other applications, eg EDI.

In summary then, Hoechst are to be truly congratulated on their highly structured and imaginative approach to the extremely critical area of an organisation's communications strategy.

10 The 1988 X.400 Recommendations — A Cause for Concern?

INTRODUCTION

It has already been stated earlier that the original CCITT X.400 recommendations, as ratified in 1984, contained a number of errors and ambiguities. Although these were effectively dealt with in the subsequent 'Implementors Guides' there was a clear need for a 'clean' set of base documents for this important topic area.

In addition to this the concepts of OSI application layer architecture, which were not fully established when the original X.400 work was being completed, had since been finalised. From this it was clear that the existing MHS architecture model and sub-layering techniques did not fit within this new framework.

Certain areas of the 1984 recommendations were left for further study at the time of ratification, which was generally due to a lack of time within the study period. Further, since the ratification of the original messaging work, a number of associated topics have been defined, such as security and directory functions, which have a bearing on this area of development. In addition to these, the requirement for a number of extra MHS facilities had also been identified, such as a message store and connection to physical delivery services.

All of the above reasons and others meant that the CCITT would have to continue its work on the message handling topic throughout the study period after the ratification of the original documents. The goals of this development were to:

— produce a set of documents which were as nearly free from errors and ambiguities as was possible;

— encompass the new concepts of the OSI application layer;

— retain backward compatibility with existing 1984 X.400 implementations;

— include some new features and facilities which were deemed to be of use.

This development process has culminated in the production of the 1988 X.400 drafts, completed at the time of writing (early 1988), which will process through an editorial phase and voting period before ratification late 1988.

The original ISO Message Oriented Text Interchange System (MOTIS) work was an adoption and expansion of the CCITT 1984 recommendations. The development of X.400 by ISO was largely to allow message-handling facilities to encompass private messaging requirements. MOTIS provided the user with the option of linking private messaging domains directly, hence yielding the possibility of forming private messaging systems, ie PRMD to PRMD interworking. In Europe it is important to take account of this work as it has formed the basis of the major European messaging profiles from CEPT and CEN/CENELEC. Since these profiles are now classified as European Pre-Norms, and will eventually reach full European Norm status, they have special significance for European procurement situations. This is now especially apparent for Government Departments and Local Authorities after the recent EEC Decision

(87/95) which will apply in all IT procurement situations. In the light of this it is perhaps surprising that MOTIS was not progressed through to full International Standard (IS) status, instead of being abandoned at the Draft-IS (DIS) stage in early 1987. This decision was taken in the light of the forthcoming 1988 work from CCITT and the amount of time and effort required to bring the original ISO documents through to full IS status. ISO preferred to discard the original work and to adopt the current drafts from the CCITT and to work with them to achieve a standard and recommendation with common text.

Although the new CCITT recommendations for message handling will only be ratified during late 1988 and subsequently published sometime in early 1989, it is likely that their adoption will be more rapid than the original work. In 1984 there was little experience of X.400 concepts and hence some inertia to overcome before their implementation. The latter half of 1987 and early part of 1988 have really been the take-off period for 1984 X.400 products and services, a time lapse between ratification and implementation of approximately four years. 1988 X.400, however, is emerging when initial experience with messaging concepts has been gained and hence there is no inertia to overcome. Further, the added value in terms of features and facilities which these new recommendations offer, will be the real incentive in the move towards their implementation. It is possible to conceive of the first products appearing on the market by late 1990 or early 1991, ie a substantially shorter time-scale. It must be emphasised that the situation with this updated recommendation in no way parallels development with others, such as X.25. The X.25 recommendations have been updated on three occasions, the latest of which was in 1984. This later version of X.25 offers no real added value in terms of features or flexibility, thus, there is a lack of incentive to implement them when earlier versions work adequately well. 1988 X.400 is a definite improvement on what 1984 X.400 can offer, hence its adoption would seem to be assured.

THE STRUCTURE OF THE 1988 X.400 RECOMMENDATIONS

The revised structure of the 1988 X.400 recommendations is as follows:

X.400 – System and Service Overview. A good introductory document which looks at the general concepts of MHSs including short sections on the many new features which have been included.

X.402 – Overall Architecture. A fairly detailed introduction to the concepts of 1988 messaging characteristics, the MHS model and how this relates to the new application layer architecture.

X.403 – Conformance Testing for 1984 X.400. This recommendation describes the test methods, criteria and notation to be used for conformance testing 1984 MHSs (as supplemented by Version 5 of the Implementor's Guides). This simply acts as a definition of what an 1988 system will interwork with.

X.407 – Abstract Service Definition Conventions. Specification of the services to be provided in an MHS environment in an abstract form. This effectively separates the description of the service from its concrete realisation.

X.408 – Encoded Information Type Conversion Rules. Similar to the 1984 recommendation but with the items classified for 'further study' now complete. It specifies the algorithms which the MHS will employ when converting between different encoded information types, eg IA5 to telex.

X.411 – Message Transfer Systems: Abstract Service Definitions and Procedures. Defines the abstract service provided by the MTS and specifies the procedures to be performed by MTAs to ensure correct distributed operation of the MTS.

X.413 – Message Store: Abstract Service Definition. This recommendation defines the procedures for using the Message Store (MS) and indirect-message submission through the MS to the MTS.

X.419 – Protocol Specifications. This details the remote UA access protocol (P3), the MS access protocol (P7) and the message transfer protocol (P1).

X.420 – Interpersonal Messaging System. This deals with the Interpersonal Messaging Service (IPMS), a user service for the exchange of messages between human users. The major difference between this recommendation and the 1984 version is the lack of the Simple Formattable Document (SFD) specification which has now been dropped.

The correlation between the CCITT MHS and ISO MOTIS numbering schemes is as follows:

CCITT	ISO
X.400	10021-1
X.402	10021-2
X.407	10021-3
X.411	10021-4
X.413	10021-5
X.419	10021-6
X.420	10021-7

Other related ISO and CCITT recommendations are as follows:

ISO 7498 : X.200 - OSI: Basic Reference Model;

ISO 8824 : X.208 - OSI: Specification of ASN1;

ISO 8649/2 : X.217 - OSI: Association Control: Service Definition;

ISO 9066/1 : X.217 - OSI: Reliable Transfer: Model and Service Definition;

ISO 9072/1 : X.219 - OSI: Remote Operations: Model, Notation and Service Definition.

THE 1988 MHS FUNCTIONAL MODEL

When compared to the 1984 MHS, the new functional model (see Figure 10.1) has basically three additional features:

— The concept of access units has been formally introduced, although these will be, or are already offered with 1984-based X.400 services, to allow MHS users to communicate with telex and telematic services. The rules for coded information conversion, which were left for further study in the first issue of the recommendations, have now been fully defined and hence it is now possible to standardise the conversion of message contents in order to achieve a coherent strategy on a worldwide basis.

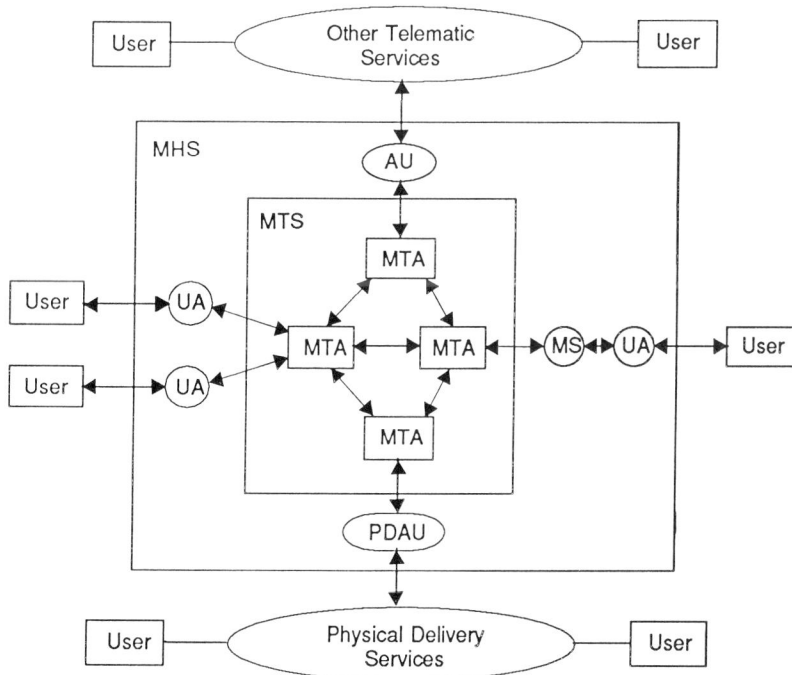

Figure 10.1   The 1988 MHS Functional Model

— Access to physical delivery (PD) services was identified as another means of achieving a critical mass situation with X.400 systems, thus it was decided to introduce this into the new recommendations. An MHS user will be able to have a message delivered to its recipient by sending the message to the PD access unit (PDAU) which will produce a hard copy together with an envelope addressed using the electronic addressing information provided by the originator. A special type of Postal O/R address has been introduced which contains the typical sort of information to be found in a normal postal address along with additional information about the type of service to be used, eg postal or courier.

— The final feature to be highlighted from the functional model is the potential for the inclusion of a message store (MS) between an MTA and its remote UA. This overcomes the problems outlined in Appendix 7 with respect to 1984 remote UA arrangements because the UA is effectively buffered from its MTA by its MS.

Architectural Changes to the 1988 MHS Model

Quite late in the 85/88 study period it was decided that the then current MHS drafts should receive a major re-draft to attempt to embody the newly established concepts of OSI Application Layer architecture. This re-draft involved the removal of 1984 sub-layering concepts and the introduction of blocks of funtionality which can be combined in a number of ways to provide different forms of connection dependent upon the operation to be accomplished.

In Appendix 4 the OSI Application Layer structure is reviewed in some detail. This indicates that communication between application processes is represented in terms of an association between application entities (AEs) residing within the application layer. Each AE is a grouping of functionality which has been combined in order to achieve specific tasks such as message handling or file transfer. The functionality of an AE is sub-divided into a set of one or more application service elements (ASEs). Any interaction between AEs is described in terms of their use of services provided by ASEs. Within the 1988 X.400 recommendations the following ASEs are described:

— Access to the Message Transfer (MT) service:

  • Message Submission Service Element (MSSE),

  • Message Delivery Service Element (MDSE),

  • Message Administration Service Element (MASE);

— Access to the Message Store (MS) service:

  • Message Submission Service Element (MSSE),

  • Message Retrieval Service Element (MRSE),

  • Message Administration Service Element (MASE).

These ASEs are concerned with the operation of message transfer/receipt and as such require the support of other ASEs which are concerned with establishment/release and information transfer in an OSI environment. Such facilities are provided by the following ASEs:

— Remote Operation Service Element (ROSE). Supporting interactive request/reply operation within the MHS model.

— Reliable Transfer Service Element (RTSE). Supporting the reliable transfer of application data.

— Association Control Service Element (ACSE). Supports the establishment and release of connections between a pair of AEs.

In order to observe how these ASEs interact refer to Figure 10.2 which shows the permissible application-contexts. There are three basic ones to consider: a remote user accessing the MTS; a user accessing the MS; and the basic message transfer operation.

MESSAGE STORE

One of the obvious omissions from the CCITT's 1984 recommendations was the lack of a message store (MS) between a MTA and its remote UAs. This meant that if the remote UA was 'off-line' for a period

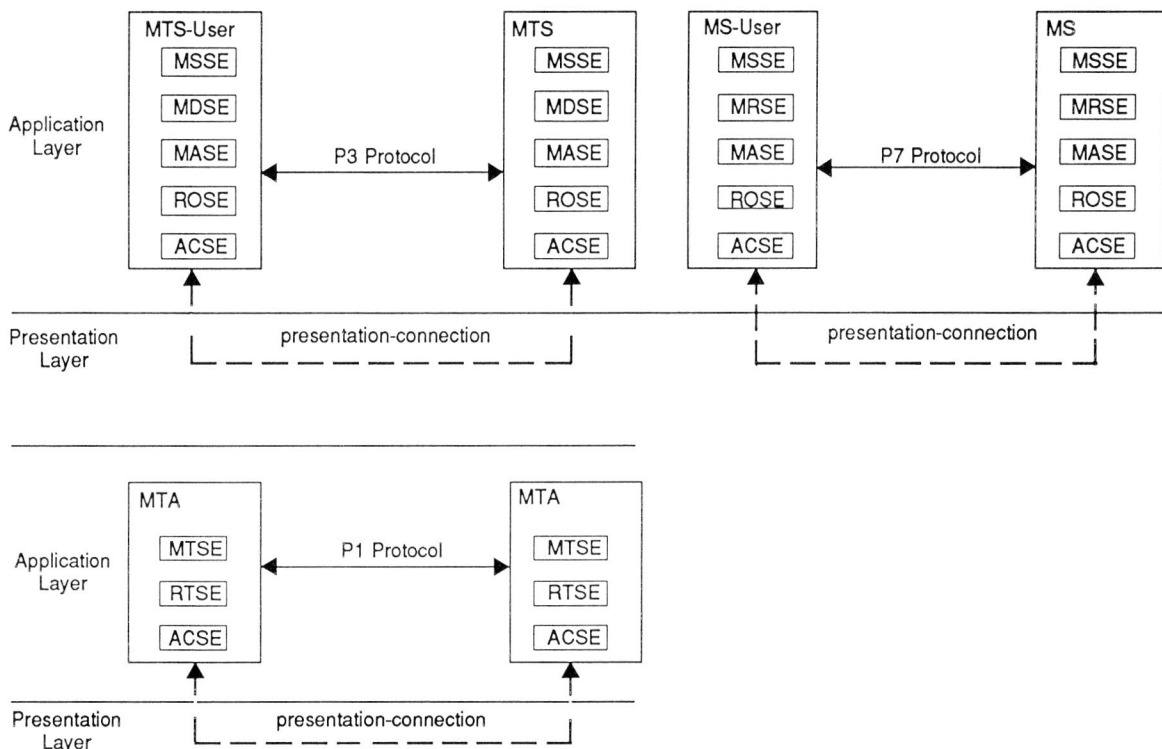

Figure 10.2   MHS Application Contexts

of time it could be 'flooded' with messages at the log-on stage. If at this time there is insufficient information storage available then message loss can result. The 1988 X.400 recommendations include MS concepts so allowing the remote UAs to be 'off-line' without the danger of message loss when the UA becomes 'live' again.

The functionality defined for the message store can be summarised as follows:

— one MS acts on behalf of one user (ie one O/R address);

— when subscribing to an MS all messages destined for the UA are delivered to the MS; when a message is delivered to an MS the role of the MTS in the transfer process is complete;

— it is possible to request an alert when a certain message arrives;

— message submission from the UA to its MTA, via the MS, is transparent;

— users are provided with general message management facilities such as selective message retrieval, delete and list.

SECURE MESSAGING

The requirement for security features within a computer-based MHS was defined as another area for the 1988 X.400 recommendations to address. The features which have been defined are certainly extensive and can be classified into three broad classes, these are:

— Originator To Recipient. These are features which are operated on an end-to-end basis and do not require the use of MTS security features.

— Message Transfer System. Some features are provided by the MTS and hence require the UA to interact with it in order to invoke them. Note that this also implies that the MTA being used is equipped with the appropriate security functionality.

— UA, MS and MTA. Some security features apply not only to the UA and MTA but also to the MS and in particular the status of certain messages held therein.

A full list of the 1988 security capabilities is contained in Figure 10.3 which is an extract from the draft text of the 1988 version of the X.400 recommendation.

Message Origin Authentication. Enables the recipient, or any MTA through which the message passes, to authenticate the identity of the originator of a message.

Report Origin Authentication. Allows the originator to authenticate the origin of a delivery/non delivery report.

Probe Origin Authentication. Enables any MTA through which the probe passes to authenticate the origin of the probe.

Proof of Delivery. Enables the originator of a message to authenticate the delivered message and its content, and the identity of the recipient(s).

Proof of Submission. Enables the originator of a message to authenticate that the message was submitted to the MTS for delivery to the originally specified recipient(s).

Secure Access Management. Provides for authentication between adjacent components, and the setting up of the security context.

Content Integrity. Enables the recipient to verify that the original content of a message has not been modified.

Content Confidentiality. Prevents the unauthorised disclosure of the content of a message to a party other than the intended recipient.

Message Flow Confidentiality. Allows the originator of a message to conceal the message flow through MHS.

Message Sequence Integrity. Allows the originator to provide to a recipient proof that the sequence of messages has been preserved.

Non Repudiation of Origin. Provides the recipient(s) of a message with proof of origin of the message and its content which will protect against any attempt by the originator to falsely deny sending the message or its content.

Non Repudiation of Delivery. Provides the originator of a message with proof of delivery of the message which will protect against any attempt by the recipient(s) to falsely deny receiving the message or its content.

Non Repudiation of Submission. Provides the originator of a message with proof of submission of the message, which will protect against any attempt by the MTS to falsely deny that the message was submitted for delivery to the originally specified recipient(s).

Message Security Labelling. Provides a capability to categorise a message, indicating its sensitivity, which determines the handling of a message in line with the security policy in force.

Figure 10.3   1988 MHS Security Capabilities

MHS DIRECTORY ACCESS

The need for a fully-integrated directory service within an MHS environment is unquestionable. It is the means by which users can find addressing information and the source of the information which the MTS will use to establish a route for a message across the MHS between the originator and recipient. It was this goal that led the CCITT to commence work on their X.500 series of Directory Service recommendations. ISO have also adopted this work (ISO 9594) because of their Global OSI addressing needs, and undertaken a joint development process with the CCITT which parallels the approach taken with the 1988 MHS/ MOTIS work.

The principles of the directory service are adequately covered in Appendix 4, and hence, do not require further explanation here. These principles have been adopted and the directory service fully integrated with the MHS in the 1988 X.400 recommendations, (see Figure 10.4). This diagram indicates that both the UA and the MTA can have directory access by having Directory User Agent (DUA) functionality built into them. This allows access to the Directory Service Agents (DSAs) which hold the MHS addressing information. A user creates a message in cooperation with a UA and when addressing information is required, it can be retrieved from this point by allowing directory access. When a message is submitted to the MTS for delivery the system will access the directory for the information required to establish its route across the MHS.

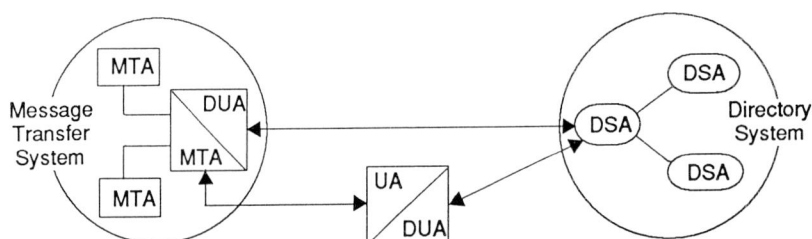

Figure 10.4   The Functional Model of MHS – Directory Interworking

SUMMARY

It is clear that the 1988 versions of the X.400 recommendations do indeed offer additional functionality. The standardisation of the message store, and the extensive security and directory access facilities means that ultimately this route will be the most useful. In addition, the backward compatibility built into 1988 will ensure that existing investments are maintained wherever possible. This is an important point, as it will be some time before implementations based on this new work will be commercially available. Hence, users should not be constrained by looking only at the implementations of the 1988 recommendations. Rather, it is better to gain experience with current 1984 implementations as the potential benefits (see Chapter 5) of these are sufficient to justify their purchase. Considering all factors it does indeed appear that the 1988 X.400 recommendations will be the future for message handling, but this does not devalue any of the current work being carried out on X.400 implementation.

11   X.400 —Towards the Future

THE CURRENT STATE OF PLAY

Today's means of electronic messaging offer insufficient scope to meet the needs of the present, let alone future, business environments. Such systems are limited either by their horizon, not being implemented in accordance with an International Standard, or the flexibility and facilities which they can provide. The user bases of these existing services have developed in isolation with the lack of a coherent strategy for their interconnection and integration. Reference to Figure 11.1 shows that by the end of 1986 there were over one million electronic messaging connections in Western Europe. From this number the two largest groups, facsimile and telex, cannot talk to each other, providing a graphic illustration of the problems mentioned so far. This situation seriously limits the benefits, in terms of the easy and flexible communications, which can be achieved by the users of these systems. In many cases they have to resort to a number of types of messaging media to enable them to address all their business and trade partners. This mode of operation is quite obviously inefficient and requires the user to be familiar with the operation of a number of systems.

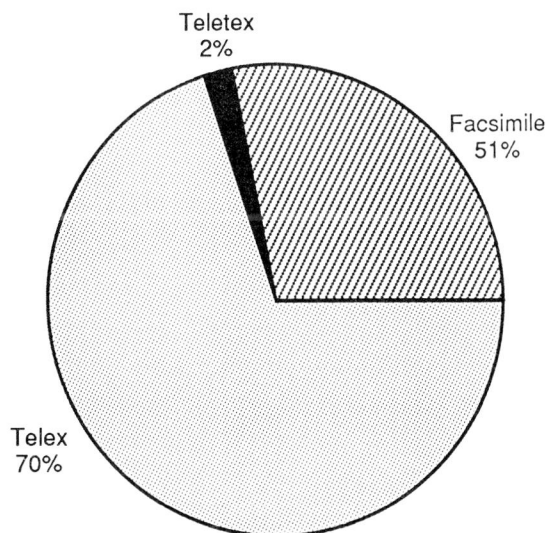

Figure 11.1   End of 1986 – 1,078,000 Electronic Messaging Connections

Source: Logica 1987 Telematic Report

So far, then, the use of electronic messaging can only be considered to be at an embryonic stage. This form of communication can potentially offer substantial advantages over the traditional telephone and postal services, but as yet it has not rivalled them seriously in terms of the volume of information carried. In order to make electronic messaging a more attractive option a substantial development of the existing infrastructure is required. Standardised access between all systems is one issue; in this way the existing user bases will be joined together and hence begin to perform as a coherent whole. Another requirement is that any new systems are developed to a recognised International Standard which will provide access to existing systems and a secure basis for all future messaging activity in its potentially varied forms. It is only by this process of development that electronic messaging will emerge from its embryonic stage to achieve the maturity it actually deserves.

The CCITT's X.400 Message Handling System (MHS) recommendations provide an internationally standardised solution by which it is possible to achieve these goals. Preservation of existing investments was always one of the main aims of this approach. In this way it allows X.400 to act as a gateway between existing systems while providing the kind of flexible base required for future systems. Hence, the horizons of X.400 systems are not limited by the lack of standardisation or flexibility. The interconnection and integration of existing systems along with new ones will mean that investments are maintained, and in fact their value should be increased substantially because of the potential to access whole new communities. Electronic messaging should eventually emerge as the major form of intra-/inter-business communications, allowing real improvements in the efficiency and speed of information transfer.

To consider systems based upon the X.400 recommendations merely as being useful for electronic mail is to seriously underestimate their capabilities. Message handling systems can transfer any type of structured or unstructured data and, hence, provide a flexible base for user specific applications. Various examples of these have been discussed in this book, but it must be emphasised that the potential for others is only limited by the imagination of the user. As their experience with X.400 develops they will start to see how it can impact on other parts of their business operation rather than just electronic mail. Ultimately it is possible to foresee a single X.400 connection, providing the user with access to many applications.

Open Systems Interconnection (OSI), the underlying architecture upon which an X.400 message handling application is based, is likely to have a significant part to play in all aspects of computer communications. The OSI framework provides the means by which data is reliably routed and transferred between user applications. Conformance to OSI standards is going to become more and more important in the future, indeed it has already assumed some importance to certain groups. A recent EEC Decision requires that all public authorities specify adherence to OSI standards in systems costing over 100,000 ECUs (European Currency Units) or approximately £70,000 at today's values. Exemption is possible on a number of counts including lack of suitable OSI products and on a cost justification basis. However, these arguments will become less and less feasible in the near future. The emergence of X.400 as the first readily available OSI-based user application is beginning to raise the awareness of the OSI communications solution and should ultimately have a significant part to play in its adoption.

A recent NCC survey of the UK local authorities attempted to assess the level of awareness within this group which is affected by the EEC Decision. The results (see Figure 11.2) show a generally low level of awareness of OSI communications, although the PSS services and private X.25 communication systems already play a significant role. It was, however, interesting to see X.400 Message Handling Systems already playing a notable part in the plans of local authorities. At present approximately 3 per cent of the sample (over 100 responses) were actually using X.400, with a further 19 per cent planning its use in the near future. This would seem to indicate that the potential of message handling systems is already influencing the user's view of OSI as a basis for future communications requirements.

One of the major attractions of OSI-based applications, such as message handling, to the user is the removal of the need for single supplier allegiance. In the past, organisations have been constrained in their choice of new products by existing equipment investments. Lack of interconnection and integration between products from different manufacturers was one of the major spurs in the development of OSI. Purchasing products conforming to an internationally standardised communication architecture removes the worries of interconnection and integration. In this way the user can make procurement decisions based purely on cost and performance criterion, a much more desirable position. The introduction of International Standards also means that vendors have to compete in an open marktet. This situation leads to increased competition between them which eventually means better products at lower prices for the user.

Already a large number of X.400 products are beginning to appear, (see Appendix 8), some of which have been exhibited at the recent spate of interworking demonstrations. These events, particularly ceBIT 87 and Telecom 87, are beginning to convey the message that X.400 has really arrived, that it is no longer just another International Standard. Both ceBIT 87 and Telecom 87 marked a significant step forward for X.400 because they demonstrated the interworking of 'real' products and services rather than just prototypes. X.400 services, the messaging platforms for individual countries, are starting to emerge. Telecom 87, in particular, illustrated the potential for the interconnection of such services on a global basis. This event has given a substantial pointer towards the first stages in the development of a worldwide X.400 messaging infrastructure. But what does the future hold?

THE FUTURE

The results of the CCITT X.400 questionnaire contained within this book (see Chapter 7) are a very important indication of the worldwide acceptance and adoption of message handling concepts. A total of 28 administrations and service providers from 25 countries indicated that they will be providing X.400 services within the next couple of years. These results give a graphic illustration of the potential for a worldwide X.400 messaging infrastructure. In addition, consider that most of these services will provide access to telex and telematic services and it becomes clear that the integration and interconnection of today's electronic messaging media is only just around the corner.

Further insight into the potential growth of X.400 can be gained from forecasts made within Volume 1 of Logica's 1987 Telematica Report (see Figure 11.3). These figures build upon the survey information detailing the numbers of electronic messaging connections within Western Europe at the end of 1986 (see Figure 10.1). By the end of 1992 the total number of electronic messaging connections within Western Europe is expected to have risen by some 200 per cent to approximately 3.5 million. This massive predicted growth is largely accounted for by two groups—facsimile and X.400 connections. Use of facsimile is expected to have risen by just less than 500 per cent since the 1986 figures, while X.400 connections, which played no part in the earlier results, are anticipated to have risen to the point were they rival telex for second most popular service. This is a very impressive indication of the potential rapid emergence of X.400 Message Handling Systems. Along with this growth comes the capability of interconnecting and integrating the existing systems and services and hence the creation of a coherent messaging infrastructure on a worldwide basis.

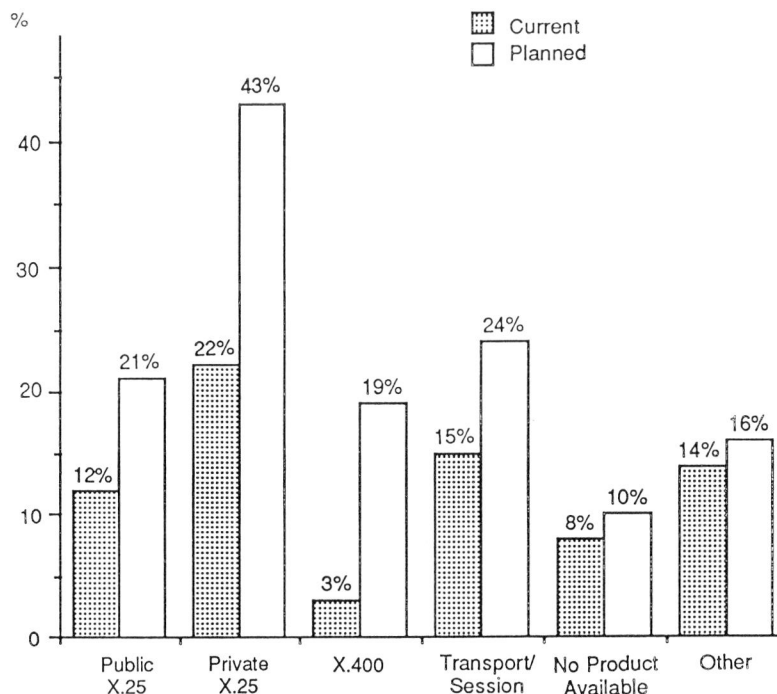

Figure 11.2   NCC Local Authority OSI Survey Results

X.400 Mailbox
23%

Facsimile
51%

Telex
24%

Teletex
2%

3496K Connections

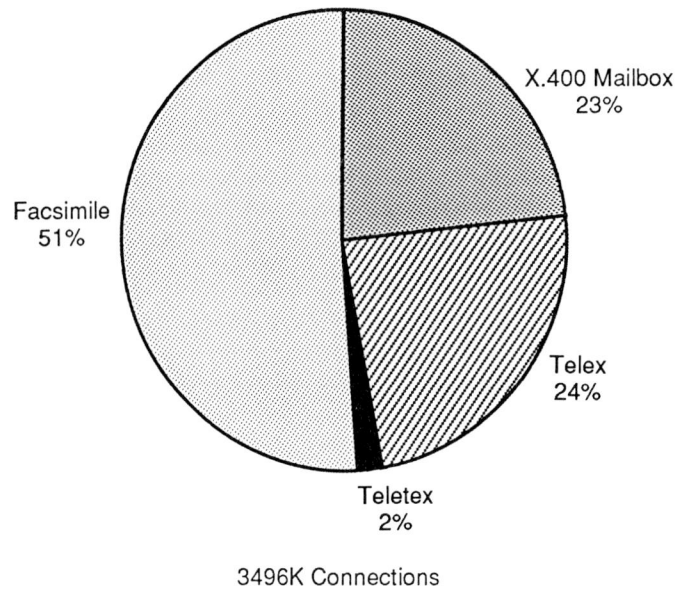

Figure 11.3   End 1992 – Forecast of Electronic Messaging Connections and X.400 Publicly Accessible Mailboxes in Western Europe

Source: Logica 1987 Telematic Report

The potential growth of X.400 systems and services is in part due to the fact that the recommendations are internationally agreed and, therefore, a stable basis for implementation. The development of the new 1988 X.400 recommendations (see Chapter 10) should not be considered as altering this situation. The 1988 recommendations are in reality a logical evolution of the earlier ones. When implementations are available, probably 1990/91, they will provide a significant upgrading of facilities to the MHS user. These additional features will include a message store, extensive security services and integrated directory access. Systems based on the 1988 recommendations will not threaten existing investments in 1984 X.400 because backward compatibility has been ensured. It is clear then that the latest version of the X.400 recommendations will have a substantial impact on the future of message handling, providing for a logical upgrading of capabilities without having to be considered as a threat to existing investments.

X.400 message handling should not be viewed as merely electronic mail, as this severly undervalues its true potential. However, it is fair to say that initially at least electronic mail will be the major application for X.400 systems and services. In the future though X.400 will offer a generic platform for many applications, some of which will have not yet even been conceived. In the future then it is possible to envisage a user working from a single X.400 terminal having access to most of the business applications which could possibly be required and being able to communicate with all other electronic messaging media. In this way X.400 Message Handling Systems will have truly become 'The Messaging and Interconnection Medium of the Future'.

Appendix 1
Glossary

This glossary defines many of the message handling terms used throughout the book; it also defines many of the commonly used terms from the general area of communication based on the principles of OSI

Abstract and transfer syntaxes These reside within the OSI presentation layer and are concerned with the specification of the structure and contents of application data. An abstract syntax will provide for the definition of application datatypes and indicate their required structure. When the transfer syntax is applied to the resultant abstract syntax it will convert it into a form suitable for transmission.

ACSE Associated Control Service Elements. Functions defined as part of the application layer, common to a number of application services (DIS 8649/2 and DIS 8650/2).

Access Unit (AU) In the context of a Message Handling System (MHS) these are the units which will act as gateways to other communication systems not based upon X.400. Examples of these would be telex and other telematic services such as facsimile and teletex.

A D Addendum to ISO Full International Standard (eg ISO 8348/ADDI).

Administration In the context of MHSs CCITT define an administration as either a PTT (Post, Telephone and Telecommunications operator) or a Recognised Operating Agency (RPOA).

Administration Management Domain (ADMD) These are management domains within the MHS which are under the control of an administration. These will provide the X.400 message handling services for users.

Application-entity The part of an application process which concerns OSI.

Application process An OSI term to describe a user of the OSI infrastructure – whether it be an application program, a human operator or a process control device.

Architecture A framework for a computer or communications system which defines its functions, interfaces and procedures.

ASCII The American Standard Code for Information Interchange. Based on the ISO 7-bit data code, usually transmitted in 8-bit characters incorporating a parity bit.

ASN.1 Abstract Syntax Notation One. Part of the OSI presentation layer standards (ISO 8824 and ISO 8825).

Asynchronous	Simple communication of data character by character which generally relies on a host screen formatting.
BAS – Basic Activity Subset	One of the defined subsets of the session layer (ISO 8326 and ISO 8327).
BCS	Basic Combined Subset. One of the defined subsets of the session layer functionality.
Bit	Binary digit. When referred to in bits per second (or bps) it indicates the transmission rate of a communications link.
Body	In the context of an MH service this is the component of a message which contains the originator information which is to be communicated.
Body part	The body of an X.400 message may consist of several constituent sections which are termed body parts.
Byte	A group of eight binary digits which form the basic binary word used to convey information. For example, the bit sequence within a byte can represent the code for a letter or a number (ie ASCII).
CASE	Common Application Service Elements. A set of OSI layer 7 standards which provide common functions for the application programs and other application layer standards (DIS 8649 and DIS 8650). No longer widely used—more likely to see ACSE and CCR.
CCR	Commitment, concurrence and Recovery. Functions defined as part of CASE which are common to many specific applications (DIS 8649/3 and DIS 8650/3).
Connectionless service	Where no permanent connection can be assumed, and no connection establishment takes place prior to communication.
Connection-oriented service	Where a permanent connection (either logical or physical) exists for the duration of the communication.
Content	In the context of an X.400 MHS this is the information within a message that the MTS will neither examine nor modify, unless conversion is specifically requested, during its transfer of the message between originator and recipient.
Content type	This is an identifier placed upon the message transfer envelope which specifically defines the type of contents.
Conversion	Message handling conversion is the situation where the MTS will perform a transformation from one encode information type to another in order that the message can be delivered into the recipient's environment.
CSMA/CD	Carrier Sense Multiple Access/Collision Detection. One of the major classes of low-level network technology, and a methof of preventing data corruption used mainly for local area networks (LANs). Ethernet is an example of this type. It is specified in OSI Standard ISO 8802/ 3, based upon the work of IEEE committee 802.3.
Delivery	When the recipient's MTA receives a message it will deliver it to the recipient's UA in a delivery envelope containing the message body and any header material.
Draft Addendum (DAD)	A draft addendum to an ISO Full International Standard.
Directory	Directory is a generic name which implies a collection of Open Systems which are co-operating to provide directory services to both

	users (eg finding a recipient's X.400 O/R names) and to systems (eg network addresses to establish the required route).
Directory name	The name of an entry within a directory.
Directory System Agent (DSA)	These are the main building blocks of an OSI directory which provide access to directory information for both DUAs and other DSAs.
Directory User Agent (DUA)	The user interface to the directory service which will employ its own DSA, and others where necessary, to locate the required directory information for the user.
Distribution list	In an MHS a distribution list is a specified list of users, and their associated address information, to whom an originator can send a message by invoking the list via its directory name.
Domain Defined Attributes (DDA)	These are X.400 O/R name attributes which are specific to the management domain within which the recipient exists. These may allow a message to be routed to systems outside the MHS.
DIS	Draft International Standard. Final stage prior to becoming a Full International Standard from ISO (eg DIS 7942).
DP	Draft Proposal. A proposed standard from ISO at the first stage of development but with some technical stability (eg DP 8613).
EN	European Norme. A standard within the European Community.
End-system	Strictly used in the OSI context to define an 'open system' that can communicate with other end-systems via OSI protocols, as distinct from a relay or gateway that performs an intermediate routeing function.
Entity	Active element within one OSI layer which employs the services of the next lower layer to communicate with a peer entity.
ENV	European pre-standards from CEN/CENELEC/CEPT.
Envelope	A message handling envelope will contain the information which the originator has specified for transfer and details to be used during the submission, relay and delivery phases of message transport across the MHS.
Explicit conversion	This is a specific form of MTS content conversion in which the originator has specified both the initial and final encoded information types.
Electronic Data Interchange (EDI)	EDI is the electronic transfer of structured business data, directly between computer systems, in accordance with agreed message standards.
Electronic Data Interchange for Administration, Commerce and Transport (EDIFACT) ISO 9735	EDIFACT is the rencently agreed ISO Full International Standard for EDI which defines the syntax to be used during such transactions.
ESPRIT	European Strategic Programme of Research in Information Technology.
Ethernet	A type of local area network based upon CSMA/CD technology—originally developed by DEC, Intel and Xerox.
EurOSInet	European demonstration of an OSI network by a number of leading vendors.
Frame	A unit of HDLC. A sequence of bits which make up a valid message,

	containing flags, control field, address field, a frame check sequence, and optionally an information field.
FTAM	File Transfer, Access and Management. One of the protocols being developed for the OSI application layer; specified in DIS 8571.
Functional standards	Identified 'stacks' of base standards to allow construction of interworking products.
Gateway	An intermediate system in the communication between two or more end-systems which are not directly linked and/or observe different protocols (eg between an OSI system and a non-OSI system).
GOSIP	Government OSI Profile. Work from the CCTA to develop functional standards for the UK.
Header	A component of an Interpersonal (IP) message which defines the control information which characterises the IP Messaging Service (IPMS).
HDLC	High-level Data Link Control. A standard for frame structures in connection with data communications protocols, at the data link layer (ISO 3309 is concerned with HDLC frame structure).
Host	A computer system on which applications can be executed and which also provides a service to connected users and devices.
IGOSINET	An OSI network organised by IGOSIS. Purpose is to encourage interworking.
IGOSIS	Implementor's Groups for OSI. Organised by ITSU of the DTI.
Interpersonal Messaging Service (IPMS)	This service provides human users with facilities to exchange messages via the MHS. It is similar in many ways to existing electronic mail facilities and allows many types of message content to be exchanged, eg from ASCII to Group 3 facsimile images.
IP-Message	The content of a message in the IPMS, ie the user information to be transferred.
Implicit conversion	This is another form of MHS content conversion in which the MTS will perform the selection of initial and final encoded information types.
IPM Protocol (P2)	The protocol which operates between IPM-User Agents for the provision of the IPMS.
IPM-User Agent	The IPM-UAs are the specific class of UAs which provide the IPMS. Further classes of UAs will eventually be defined to meet other specific user requirements.
Interconnection	A term often used to define a lesser level than full interworking such that two computer systems can communicate and exchange data but without consideration of how the dialogue between application processes is controlled or how the data is presented and recognised.
Interworking	Ultimately, the achievement of proper and effective communication or 'linking' between different application processes or programs and data, may be on different systems from different manufacturers, remote from each other and connected by some transmission medium or network.
IS	International Standard. Fully agreed and published ISO standard (eg ISO 7498).
Integrated Services Digital Network (ISDN)	ISDN services will emerge out of the current modernisation, from analogue to digital, of the world's telephone networks. They can

provide integrated voice and data facilities by extending the digital connection right to the customer's premises. The CCITT are currently developing their I-series of recommendations for ISDN which will be OSI-compatible at the lower three layers of the model.

IT	Information Technology. A term used to encompass the methods and techniques used in information handling and retrieval by automatic means, including computing, telecommunications and office systems.
JTM	Job Transfer and Manipulation. One of the protocols being developed for the OSI application layer for activating and controlling remote processing (DP 8831 and 8832).
Kernal	Service elements within the session layer which are necessary to set up and close down a connection; part of ISO 8326 and ISO 8327. Also used to describe basic elements of CASE (DIS 8649 and DIS 8650).
LAN	Local Area Network. Spans a limited georgraphical area (usually a building or a site) and interconnects a variety of computers and other devices, usually at very high data rates.
LAPB	Link Access Procedures Balanced. A variant of HDLC used between peer systems, which is the basis for layer two of X.25 (as an example of standards in this area, ISO 7776 is concerned with X.25 LAPB compatible DTE Data Link Procedure).
Layer	In the OSI Reference Model, used to define a discrete level of function within a communication context with a defined service interface—alternative protocols for a particular layer should then be interchangeable without impact on adjoining layers.
MAP	Manufacturing Automation Protocol. Initiated by General Motors in order to force suppliers to adhere to a prescribed set of OSI-based standards.
Medium	The physical component of a network that interlinks devices and provides the pathway over which data can be conveyed. Examples include coaxial cable and optical cable.
Message Handling Systems (MHS)	The generic term used to describe an X.400 message handling environment which is a collection of interconnected MTAs and UAs.
Message Oriented Text Interchange Systems (MOTIS)	The ISO name for both the message handling environment and their draft messaging standards, a superset of X.400 expanding the message handling functionality into the private domain. The original MOTIS work based upon the 1984 X.400 recommendations was recently abandoned but the name has been resurrected for their joint work with the CCITT on the 1988 version of message handling.
Message Transfer Agents (MTAs)	These are the main store-and-forward building blocks of an MHS; they provide the message transfer and relay capabilities which are used by the UAs.
Message Transfer (MT)	This term is used to describe the store-and-forward message delivery process which is employed within an MHS; this movement of the message is between the MTAs of the system.
Management Domain (MD)	In order that the whole message handling environment can be subdivided into sections for administrative and maintenance purposes the concept of management domains (MDs) has been included within the X.400 recommendations. There are two types of MD which have been defined in X.400: ADMD; PRMD.

Management Domain Name

In order to enable systems to route messages within the MHS each MD has a name which can be quoted in the address field of a message.

Message Transfer Layer (MTL)

This concept is strictly 1984 X.400 as it is not in line with current concepts of the OSI application layer. The MTL is the lower of two sublayers within the application layer which provides the MT service, the upper of these two layers is the User Agent Layer or UAL.

Message Transfer (MT) Service

The MT service is that service within the MHS which provides the basic store-and-forward message transfer capability. In general it is not concerned with the contents of a message unless a content conversion is to be performed.

Message Transfer Protocol (P1)

P1 is the protocol which provides the MT service by enveloping the content and header before it is transferred across the message handling environment.

Message Transfer System (MTS)

The MTS is the interconnection of all MTAs which provide the message transfer service elements.

Message Store (MS)

This is a new concept from the 1988 X.400 recommendations which provides an intermediary message store 'buffer' between a remote UA and its MTA.

Message

In the context of an MHS this the unit of information transferred by the MTS and consists of both the envelope and its contents.

Multiplexing

The carrying of more than one data stream over the same connection (apparently) simultaneously.

Network

A collection of equipment and/or transmission facilities for communication between computer systems (whether a single dedicated link, line, dial-up PSTN (telephone) line, public or private data network (PDN), satellite link, etc). More correctly in the OSI context used to define the achievement of end-to-end communication between end-systems, however accomplished.

Network architecture

A generic term for the layered approach which individual vendors take towards development of their communications and applications products. Examples include IBM, and SNA, DEC and DNA, Honeywell and DSA, and ICL and IPA.

Network layer

Level three of the OSI model. It is the means of establishing connections across a network such that it then becomes possible for transport entities to communicate.

Node

A focal point within a network at which information about a network entity is considered to be located.

NSAP

Network Service Access Point. The service access point which allows entities within network and transport layers to interact. It is situated upon the network layer boundary and located by its address, which is the subject of ISO 8348/ADD2.

ODA

Office Document Architecture. A proposed architectural structure for an office document which allows its logical and layout structure to be defined in an unambiguous manner.

ODIF

Office Document Interchange Format. A structure for interchange of complex office documents.

ONA

Open Network Architecture. A set of OSI profiles from British Telecom which are to be supported when connected to BT services.

OSI

Open Systems Interconnection. A term which is used to describe the

	area of work concerned with vendor independent standardisation; largely carried out under the guidance of ISO.
OSI gateway	A method of providing access to an OSI network from a non-OSI system by mapping the sets of protocols together.
OSI reference model	Seven layer model defined by an ISO subcommittee as a framework around which an Open Systems Architecture can be built. It describes the conceptual structure of systems which are to communicate.
Originator/Recipient (O/R) Address	A descriptive name for a UA which contains certain characteristics which help the MTS in establishing a route to the recipients.
Originator/Recipient (O/R) Name	A purely descriptive name for a UA.
Originator	A user which may be a human being or a computer process from which the MTS accepts the message for transfer to the recipient.
Originating UA	The originator's interface to the MTS. The originating UA will employ the MTS to route the message to the recipient.
Octet	This is the name used in most International Standards to indicate a byte, 'oct' implying eight, ie eight bits. As a rule octet=byte or vice versa.
Private Management Domain (PRMD)	An MHS which is outside the control of a service provider, under private control for its administration and maintenance.
Physical delivery	Physical delivery of messages originated within the MHS to users via postal, courier or other services. This service has been defined within the 1988 X.400 recommendations.
Physical Delivery Access Unit (PDAU)	The means by which an MHS user can access a physical delivery service.
Postal O/R Address	In the context of message handling a specific form of O/R address which has the characteristics of a normal postal address. In addition, it will also identify the specific physical delivery service which is to be used (ie either postal or courier).
P1, P2, P3, P7	Different classes of protocol specified within the CCITT X.400 (MHS) standards.
Packet-switching	A type of data network based upon the CCITT X.25 Recommendation, whereby a 'virtual call' is established, but individual data 'packets' may be routed across separate physical links through the network. (British Telecom's PSS is of this type).
PAD	Packet Assembler/Dissassembler. Converts data at a terminal into 'packets' (discrete quantities) for transmission over a communications line and set up and addresses calls to another PAD (or system with equivalent functionality). It permits terminals which cannot otherwise connect directly to a packet switched network to access such networks.
PCI	Protocol Control Information. Control information passed between peer entities to co-ordinate the transfer of user data. It is added to the service data unit to create a protocol data unit.
PDU	Protocol Data Unit. Created at a given layer in the stack by taking the service data unit from the layer above and adding PCI. This is the information which is passed to the peer entity.
Peer entity	Active element within an OSI layer which corresponds to an equivalent element in the corresponding layer of a different end system.

Physical layer	This is the first level of the OSI Reference Model, responsible for transmitting bit streams between data link entities across physical connections.
Presentation layer	Level 6 in the model; responsible for agreement on how information is represented.
Protocol	A set of rules for interaction of two or more parties engaged in data transmission or communication. In OSI terms interaction between two layers of the same status in different systems.
Protocol stack	The set of OSI protocols at all seven layers required for a particular function or implemented in a particular system.
PSDN	Public Switched Data Network; CCITT term for public packet switched network.
PSE	Packet Switching Exchange. A switching computer which adheres to X.25 packet-level procedures. Used by BT to describe PSS exchanges.
PSS	Packet Switch Stream; British Telecom's packet switched public data network.
PSTN	Public Switched Telephone Network.
PTT	National postal, telephone and telegraphy organisation.
Relay	A term used for a system which performs an intermediate function in the communication between two or more end-systems (eg a node in a public switched network). In the context of X.400 relay implies the movement of messages between MTAs.
Receipt	An X.400 message is only classified as being received when the user has actually accessed the message. Delivery of the message, by the MTA, to the recipient's UA does not constitute receipt.
Recipient	The intended recipient of an X.400 message is the person or system which is specified by the O/R name or address placed on the message by the originator.
Routeing	Function within a layer to translate title or address of an entity into a path through which the entity can be reached.
SAP	Service Access Point. Allows entities within adjacent layers to interact (see NSAP).
SASE	Specific Application Service Elements. Those parts of the OSI application layer which include FTAM, JTM, VT and MOTIS.
SC	Subcommittee. Within ISO, SC21 has responsibility for development of standards for OSI layers 5 to 7, SC6 for layers 1 to 4, SC18 for message handling systems.
Service	The interface between a layer and the next higher layer (in the same system) ie the features of that layer (and below) which are available for selection and the conditions reported.
Submission	When a message has been prepared the originating UA will submit it to the MTS for delivery to the intended recipient.
Simple Formattable Document (SFD)	Recommendation X.420 from the X.400 series contains a specification for defining the attributes of a document in an unambiguous manner. SFD, as it is known, was devised to allow the interchange of documents which can be described as formattable but not yet formatted.

Submission Delivery Entity (SDE)	An entity which is co-located with a remote UA, but residing in the MTL. It provides the remote UA with interactive access to the facilities of the MT service.
SDE Protocol P3	The interactive protocol which is used between an SDE and its MTA.
Session layer	Fifth layer in the model, responsible for managing and co-ordinating the dialogue between end-systems.
SNA	IBM's proprietary Systems Network Architecture, which is layered but at present has only some architectural similarity to OSI.
SPAG	Standards Promotion and Application Group. A consortium of European suppliers developing functional standards.
TC	Technical Committee. Within ISO, TC97 has responsibility for SC6, SC18 and SC2, which are the primary subcommittees developing OSI standards.
Teletex	An international service for document interchange which provides rapid exchange of text via the telephone network and other public data networks. Unlike telex, teletex is a method rather than a specific network or system (CCITT F.200, T.60, T.61 and T.62).
Trade Data Element Directory (TDED – ISO 7372)	A list of the standardised data elements to be used during EDI transactions. Soon to be updated and expanded to encompass the new applications intended for the EDIFACT (ISO 9735) EDI syntax rules.
TOP	Technical and Office Protocols. A set of functional standards designed for the office environment, initiated by Boeing in the US.
Transport classed	The method by which the options of the transport layer are grouped into five subsets.
Transport layer	Fourth level of the Reference Model, charged with guaranteeing end-to-end communication between end-systems.
Triple X	The CCITT recommendations X.3, X.28 and X.29 which jointly define standards for asynchronous terminals to access a mainframe (or X.25 packet terminal) via a PAD.
User	The functional object (eg a human user, computer process) which is using the facilities of the MHS.
User Agent (UA)	A UA provides the user with an interface to the facilities of the MTS. It may also provide local facilities such as document filing/retrieval and editing functionality, but these are outside the realm of X.400.
User Agent Layer (UAL)	This, like the MTL, is strictly a 1984 X.400 concept which does not agree with current OSI application layer ideas. The UAL is the upper of the two sublayers within the application layer which provides the user services such as IPMS.
Virtual circuit	A logical transmission path through an X.25 packet switched network established by the exchange of set-up messages between two DTEs. The circuit may use more than one physical circuit, or share a physical circuit with other virtual circuits.
WAN	A Wide-Area Network. Makes use of communications facilities which can carry data to remote sites. Could be a public data network (PDN) such as BT's PSS or a private network.
X.3, X.28, X.29	The set of Triple X protocols.
X.21	The CCITT Recommendation defining interfaces for synchronous transmission over Public Data Networks (circuit switched networks).

X.25   The CCITT Recommendation defining interfaces to packet-mode terminals on packet-switched networks, as used by British Telecom's PSS and many other national and private networks.

X.25 (1980), X.25 (1984)   The variants of X.25 agreed by CCITT at its plenary meetings in 1980 and 1984, respectively. The 1980 version is a subset of the 1984 version.

X.400   The CCITT series of Message Handling Service Recommendations.

Appendix 2
Bibliography

Bird J, Chilton P, *Text Communication – The Choices,* NCC Publications, 1988

CCITT X.400 Message Handling System Recommendations, *The 1984 Red Book*

CCITT X.400 Message Handling System Recommendations, *The 1988 Blue Book*

Gaskell P, *Migrating to OSI,* NCC Publications, 1987

ISO DIS 10021 pt 1-7, Message Oriented Text Interchange System *(MOTIS)*

ISO IS 8613, Office Document Architecture

ISO IS 8824, Abstract Syntax Notation One (ASN.1)

ISO IS 9735, EDIFACT

Price Waterhouse Information Technology Review 1987/88, Price Waterhouse, 1988

Pritchard J A T, Wilson P A, *Planning Office Automation Electronic Messaging Systems,* NCC Publications, 1982

Appendix 3
International Standardisation Activity

Looking into the world of international communication standardisation, especially for the first time, can be a confusing and fruitless task. There are a wide range of organisations operating from various countries who appear, on occasions, to be working in the same areas and hence duplicating work. The titles of these organisations are sometimes long and unmanageable and so are shortened to acronyms which, for the newcomer, often serve only to complicate the situation further.

THE MAIN STANDARDS MAKING BODIES

This initial shock of finding such confusion surrounding an activity which is designed to bring clarity to the world of computer communications does not encourage faith in the belief that anything fruitful will ever emerge. The aim of this part of the book is to attempt to throw light on this confusing situation by introducing some of the major organisations involved in the standards making process and gives some insight as to their current work on International Communication Standards. Figure A.3.3 shows the relationship between these groups and other contributory bodies.

International Standards Organisation (ISO)

ISO is an agency of the United Nations (UN) which was established in 1946 with voluntary member bodies, currently over 70, who can be either participating (voting) or observing (non-voting). ISO has the prime, but not exclusive, responsibility for international standardisation. Two important stages in the progression of a standard through the ISO machinery are 'draft proposal' (DP), when a standard has passed the formative stages and is given the number (eg DP 9999) under which it should eventually be published; and 'Draft International Standard' (DIS), when a draft has been accepted by a vote of the committee responsible for producing it. The DIS document will then be put forward for wider international vote by the members of ISO—the national standards bodies. ISO standards are usually subsequently approved by these bodies and reprinted as national standards.

The ISO structure basically consists of Technical Committees (TCs) with Sub-Committees (SCs) and Working Groups being subordinate respectively. The main TC of interest to readers of this report is probably JTC1 (see Figure A.3.1), Information Processing Systems. JTC1 or Joint Technical Committee 1 is a joint ISO-IEC committee which was formed to overcome some areas of overlap between the work of these groups. The SCs of interest are also detailed in Figure A.3.1. (*Note:* SC18—WG4 deals with the MOTIS work.)

Comité Consultatif International de Télégraphie et Téléphonie (CCITT)

The CCITT is part of the International Telecommunications Union (ITU) and is a treaty organisation who members, mainly the service providers and PTTs (ie Post, Telegraph and Telephone authorities) of member countries, have to sign a convention to join. The CCITT's major role is the harmonisation of

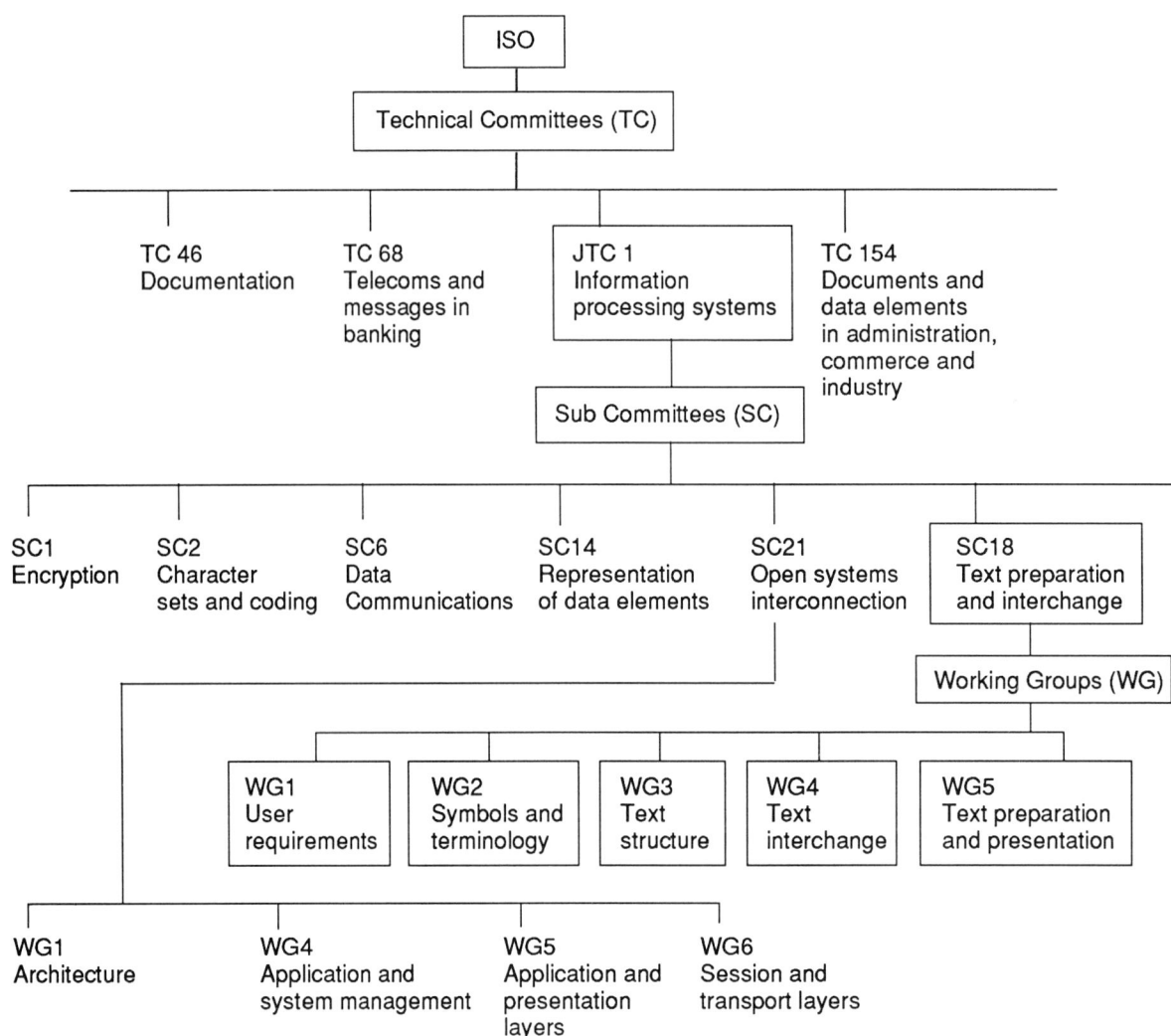

Figure A.3.1   ISO Hierarchy

communications across the world. Every four years CCITT publishes a series of recommendations on various communication standardisation topics (one of which is X.400), each of these supersedes any previous issues on the same topic. These recommendations are binding on the member countries in the realm of international communications.

Figure A.3.2 shows the CCITT hierarchy, which consists of Study Groups (SGs) and Working Parties (WPs). The work on the messaging topic falls under SG7, WP5.

International Electrotechnical Commission (IEC)

The IEC is quite an old organisation, established in 1906, which has a voluntary membership of national elements called 'national committees'. In the past there were areas of overlap between ISO and IEC but these have now been eliminated and the IEC now concentrates on standards addressing product safety and environment.

American National Standards Institute (ANSI)

ANSI is the US agent and voting member of ISO. It is a private and voluntary organisation supported by its membership, which includes representatives of manufacturers, research groups, standards groups and other interested paying parties. The organisation is supported by membership dues and document sales.

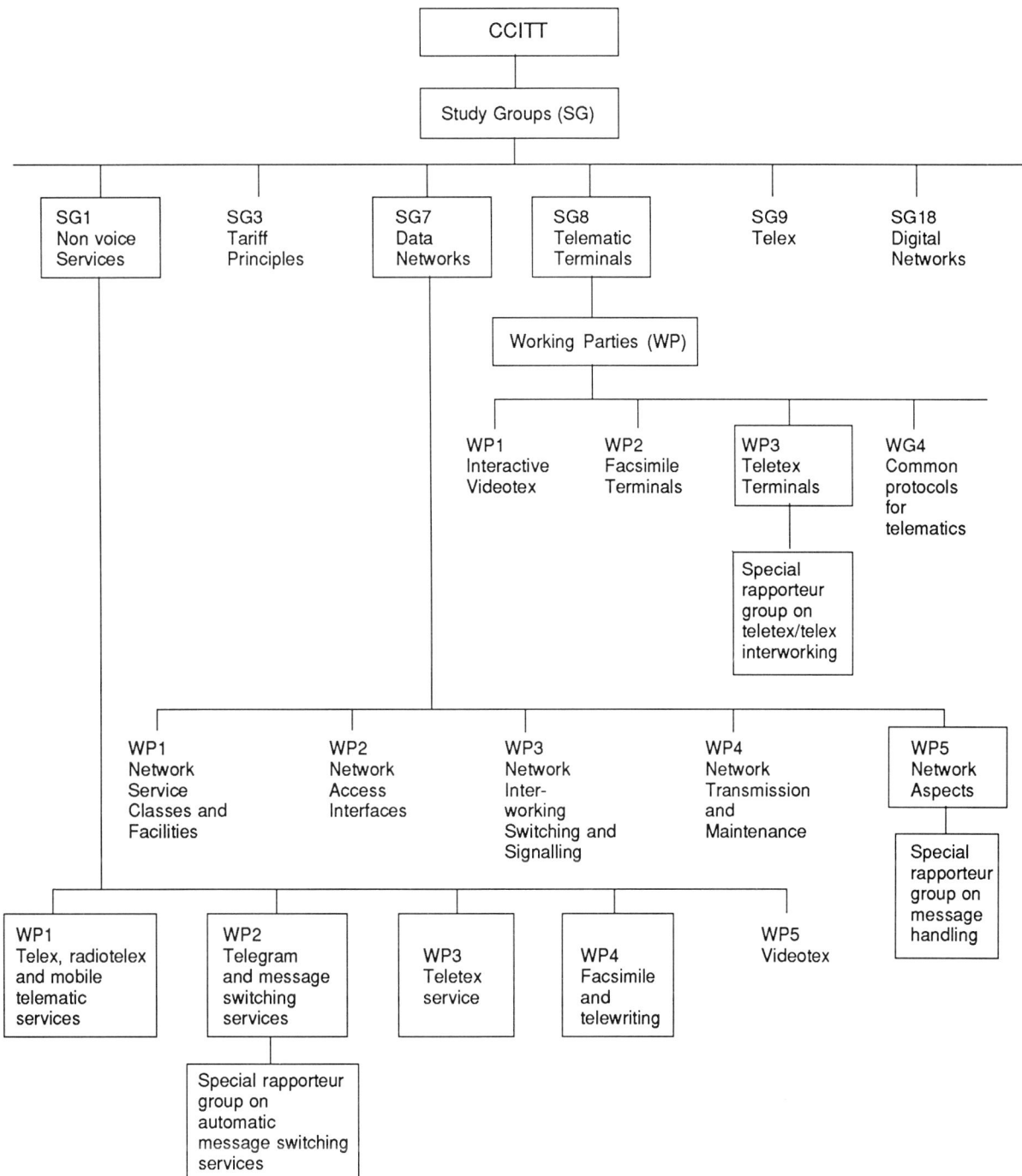

Figure A.3.2   CCITT Hierarchy

National Institute of Standards for Telecommunications (NIST)

The NIST is an agency within the US Department of Commerce and is funded by the federal government to develop procurement standards. NIST has responsibility to act as an investigatory, formative, approval and publishing organisation for standards in the federal government. Despite its narrow target, the combination of all the main roles (backed up by a very substantial budget) means that it can exercise a degree of project management which is impracticable where responsibilities are split. As a result it is highly influential in the much wider circle of national and international computing standards. NIST has strong links with other national organisations such as ANSI and IEEE.

EIA and IEEE

Electronic Industries Association (EIA) and the Institute of Electrical and Electronic Engineers (IEEE) are professional societies in the United States. Both of these organisations have contributed to the work on OSI particularly with standards for the lower layers of the reference model.

European Computer Manufacturers Association (ECMA)

ECMA is a group which consists of about 20 major manufacturers. It concerns itself primarily with computing standards, but carries out none of the other functions of a Trade Association. Its committees are open to users as observers, but only its member manufacturers have a vote at the approval stage. ECMA serves mainly as a non-voting contributor to both ISO and the CCITT and intends to support the protocols adopted by ISO for OSI.

Commission for the European Communities (CEC)

The commission acts as the civil service for the European Economic Community (EEC). The Commissioner for Industrial Affairs is responsible for telecommunications and information technology policy.

La Conférence Européene des Administrations des Postes et des Télécommunications (CEPT)

The CEPT is a conference of the posts and telecommunication authorities (PTTs) for Western Europe and certain European economic organisations.

British Standards Institute (BSI)

BSI is the UK national standards body who make contributions to ISO work and have a voting status, as the UK representative. They also publish ratified ISO standards.

Association Française de Normalisation (AFNOR)

AFNOR is the French equivalent of the BSI in the UK.

As was stated earlier, there is considerable overlap in the work carried out in European standards organisations. Efforts have been made to attempt to minimise the occurrence of such problems which have resulted in ISO and CCITT groups, who are working in the same area, holding regular liaison meetings to ensure that work is progressing in the direction that is most constructive for everyone concerned.

Figure A.3.3 indicates the general structure of the world standardisation process. The groups like ECMA and other industrial committees, along with national standards bodies such as BSI and ANSI, contribute to the actual world standards making process that revolves around ISO and CCITT.

STANDARDS BODIES INVOLVED WITH OSI

The following is a list of the standards making bodies who are involved with work on the topic of Open Systems Interconnection (OSI). The list will act as a useful reference for the reader as the names of the organisations are paired with their more often quoted acronyms.

AFNOR = Association Française de Normalisation

ANSI = American National Standards Institute

BSI = British Standards Institute

CCITT = International Telephone and Telegraph Consultative Committee

CEC = Commission for the European Communities

CEN = Comité Européen de Normalisation

CENELEC = Comité Européen de Normalisation Electronique

CEPT = European Conference of Postal and Telecommunication Administrations

COS = Corporation for Open Systems

ECMA = European Computer Manufacturers Association

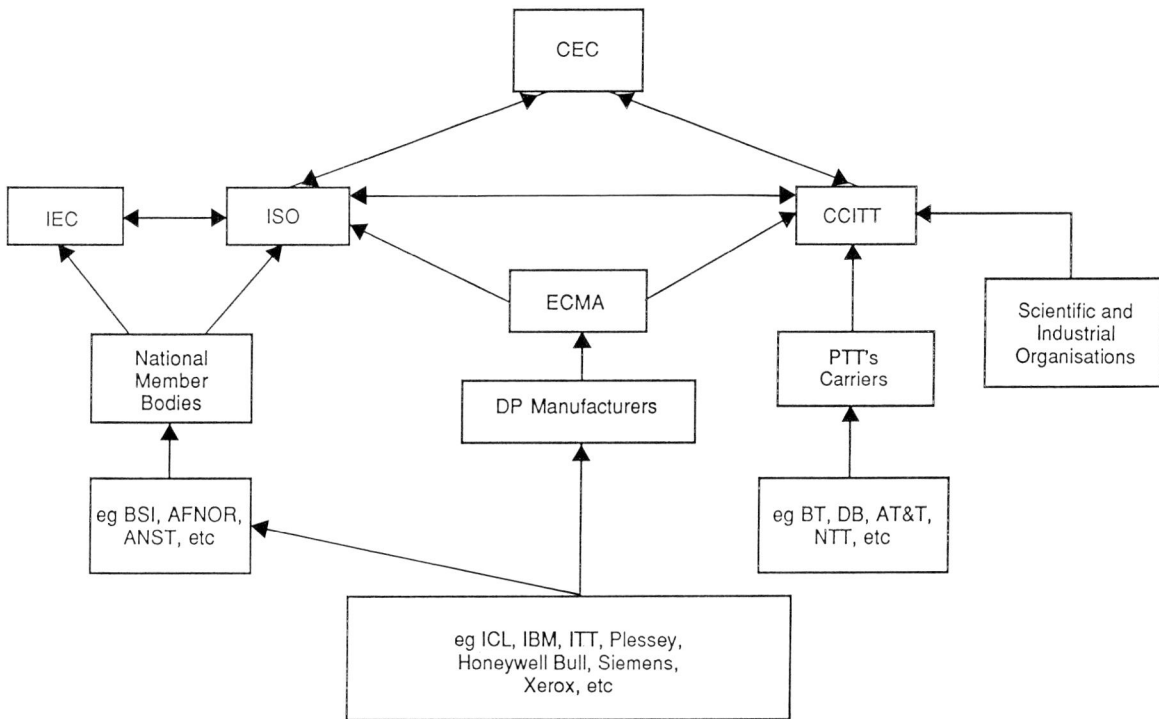

Figure A.3.3   Interactions Between Standards Bodies

EIA = Electronic Industries Association

IEEE = Institute of Electrical and Electronic Engineers

IEC = International Electrotechnical Commission

ISO = International Standards Organisation

ITAEGS = IT Ad-hoc Expert Group on Standards

MAP = Manufacturing Automation Protocol (work led by General Motors)

NBS = National Bureau of Standards

SPAG = Standards Promotion and Application Group

TOP = Technical Office Protocol (work lead by Boeing)

UK GOSIP = UK Government OSI Profile

USA GOSIP = USA Government OSI Procurement

Appendix 4

The Concepts of Open Systems Interconnection (OSI)

The interconnection/interworking of office automation and other computer-based equipment has been a problem for the user for some time. In the short term proprietary communication and equipment standards have gone some way towards alleviating the problems within single vendor domains. However, when it is required to interconnect equipment from various suppliers the problems become glaringly obvious. Some vendors have offered solutions to specific interconnection problems, but these generally have to be tackled on a one-off basis leading to complex and possibly expensive solutions which do not always fulfil the users' requirements. An example of the typical interconnection scenarios that can arise between today's incompatible hardware is shown in Figure A.4.1.

The problem outlined so far indicates that there has been a requirement for some time to define a standard interface between all computer-based communicating systems. If this were to be accomplished it would allow the user to achieve the interconnection of systems regardless of where the products have been sourced from. In this way users would be able to achieve the vendor independence for which they all crave.

ADVANTAGES OF OPEN SYSTEMS

With these problems in mind the International Standards Organisation (ISO) decided in 1977 that they

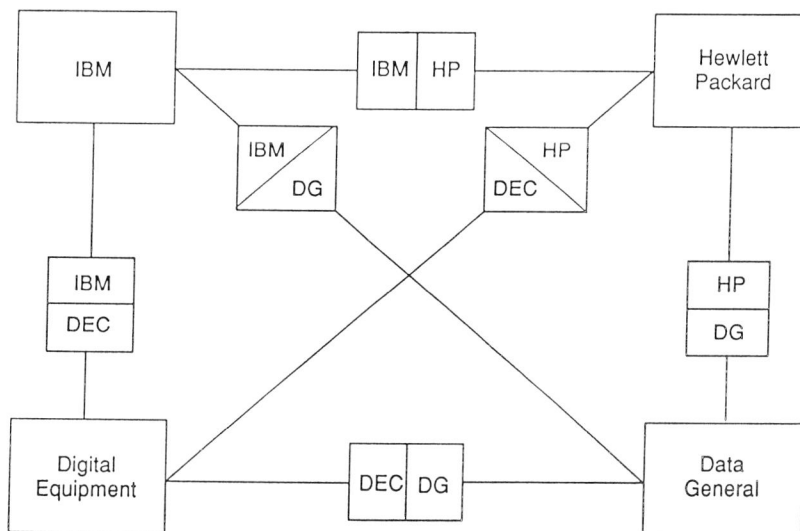

Figure A.4.1   Typical Interconnection Scenario Between Today's Incompatible Hardware

would assume a guiding role in the development of a framework for the interconnection of systems. The generic title for this work is Open Systems Interconnection (OSI) and its main principle is that all computers, regardless of make and model, will be able to talk to each other. Both ISO and the CCITT (Comité Consultatif International de Télégraphie et Téléphonie) have developed a series of protocol standards with which Open Systems (systems designed to these standards) will communicate. These protocols are based around a seven-layered architectural model.

The ultimate goal for Open Systems users is to be able to choose their equipment on the basis of its merit for a particular application without being restricted to systems from a particular supplier. Difficulties with the interconnection of equipment sourced from different suppliers should not exist as they will all be designed to the same standard/s. Hence, solutions to specific interconnection problems will be redundant, which implies saving both in terms of effort and cost. In Figure A.4.2 the interconnection scenario of Figure A.4.1 is repeated but with the use of OSI protocols for the connections between the systems. The simplification is instantly apparent.

The above are just some of the benefits of Open Systems to the user, but the only way that such systems will become available is when the equipment vendors adopt these standards. The question that has to be answered is what incentive is there for implementing these standards? Once the initial traumatic changeover period from proprietary to OSI standards—has been overcome the system vendors will start to enjoy some of the benefits of OSI. Because there are no longer any single vendor domains each company will be able to sell its products to a far wider market than before. This will lead to competition between the manufacturers to produce the best products at the most attractive prices, and because the OSI architecture already exists there is no wasted effort expended in the development of new product architectures. In line with this, vendors will be subject to fewer development risks because new equipment will be developed to internationally recognised standards.

In summary then the advantages of Open Systems are:

To the user:

— all products conform to the same standards, hence a greater choice;

— no single vendor domains;

— product purchases will be based on their merit for the application rather than whether they interface to existing equipment;

— existing equipment investments will be guarded because Open Systems will be designed to build in a modular manner;

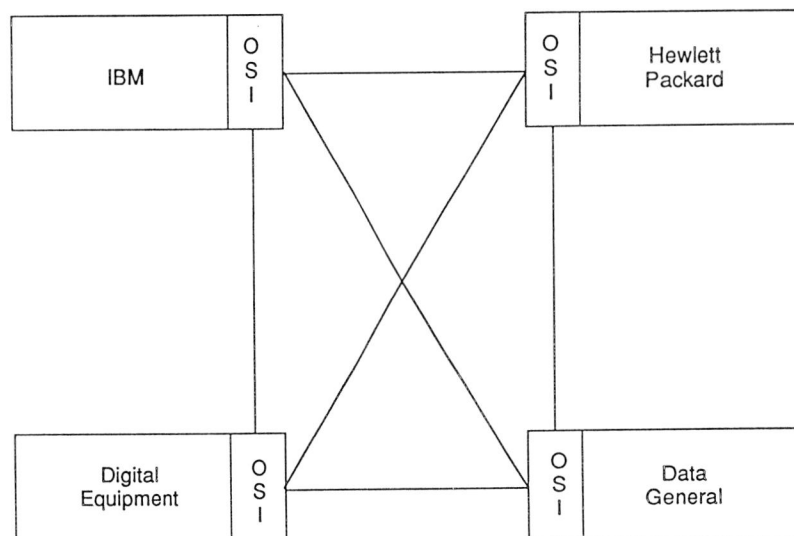

Figure A.4.2   Interconnection Utilising OSI Protocols

— reduction in procurement costs because system compatibility no longer a problem;

— no requirement for expensive and complex gateways for systems interconnection;

— improvements in communication will lead to increased distribution of, and access to, information.

To the vendor:

— wider markets due to the lack of single vendor domains;

— greater competition to produce the best products at the most attractive prices;

— no effort wasted in the development of new product architectures;

— fewer development risks;

— wider choice of experienced development staff.

BASIC CONCEPTS OF THE ISO OSI REFERENCE MODEL

ISOs first priority was to devise a standard architecture (called the Reference Model for Open Systems Interconnection currently defined in ISO IS 7498) which could be used to provide a basis for the development of the standards needed, whilst allowing existing standards to be accommodated where relevant. It also provided a method of dividing the work into manageable portions so that separate teams could work on specific protocols and services.

In practice, the reference model has been used by some computer manufacturers to provide the framework for their proprietary network architectures. The ISO reference model naturally reached a fairly stable state before any standard high-level protocols or services were defined. Thus, computer manufacturers decided that it was sensible to use it and to place their own proprietary protocols within its framework, with a view to replacing these where necessary by the standard versions when they are agreed.

Since layering has proved to be a successful technique for computing and communications functions, the ISO reference model was designed as a layered structure consisting of seven functional layers and a physical transmission medium (Figure A.4.3). Each layer is self-contained and only the interfaces to adjacent layers and the service it provides need to be fixed. Each layer uses services provided by lower layers in order to carry out its functions. Functions within a layer but in separate systems communicate with each other using protocols appropriate to that layer. No protocol acts across layer boundaries.

For each layer within the reference model there are two types of standards which are defined in order to specify its operation. These standards address two specific aspects of the service provided, namely:

— the functionality within a specific layer, ie the service provider to the layer above;

— the protocols which are utilised within a layer and between peer layers.

At the lower layers many of these standards are complete and many vendors have versions of them incorporated within their products. At the higher layers work is progressing to the extent where implementations of groups of standards are now beginning to appear.

Within OSI operation there are two basic modes which are supported:

— Connection mode. Traditional computer communications mode with a sign-on, information transfer and sign-off periods during a session.

— Connectionless mode. With this mode there is no connection established at the time of transmission. Data is passed to the network with sufficient information to enable a message to be routed to the appropriate destination.

Many of the functions contained within the layers will be applicable to both of these modes of operation, while others will only support either of the modes.

PRINCIPLES OF THE LAYERED ARCHITECTURE

The functions within layers are collected into groups which are called entities. These functional groupings can refer to different modes of operation, ie connection and connectionless. Communication between peer

Main features of layers

Peer-to-peer protocols

Application Layer →		Human users, application programs, user services, etc.
Presentation Layer →		Manipulation and conversion of structured data.
Session Layer →		Binding and unbinding of applications. Co-ordination and synchronisation of dialogue.
Transport Layer →		Control of data transport from end system to end system.
Network Layer →		Routeing and switching.
Data Link Layer →		Reduce errors introduced by physical media.
Physical Layer →		Provide means to control physical circuits.

physical media for interconnection

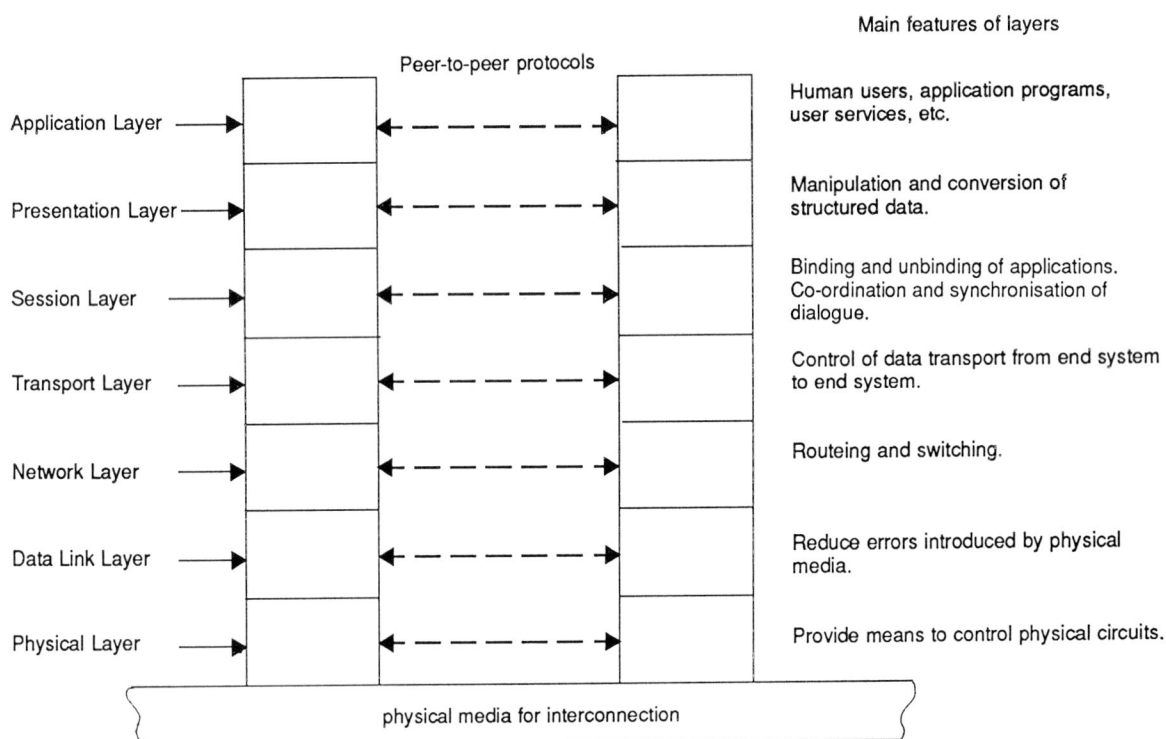

Figure A.4.3   ISO Seven-Layer Reference Model

entities, both in the same system and in separate systems, is via one or more protocols. The choice of which protocol to use within a layer will be decided by the layer itself influenced by factors such as:

(a) the quality of service required by the layer above;

(b) protocol negotiation between peer layers in communicating systems;

(c) in the case of connectionless mode the previously agreed choice of protocol (ie the most suitable choice) between peer layers in the communicating systems.

The following discussion employs the example of an arbitrary layer, N, within a communicating layer stack (see Figure A.4.4).

Entities within a layer-(N) provide a service to the entities in the layer above, ie N+1. In some cases an (N)-entity may require the co-operation of another (N)-entity, in order to provide a service to an (N+1)-entity. In this situation entities will utilise an (N)-protocol and an (N-1)-connection. The information exchanged between these (N) peer entities is separated into two parts:

(a) (N)-service-data-unit (N-SDU)—this is the data which (N)-entities require to carry out the functions requested by the (N+1)-entity;

(b) (N)-protocol-control-information (N-PCI)—this protocol information is used by the (N)-entities to co-ordinate their communication.

These two parts of the data are combined into a (N)-protocol-data-unit (N-PDU).

$$(N)\text{-SDU} + (N)\text{-PCI} = (N)\text{-PDU}$$

The data within the PDU may also be separated into smaller pieces, termed 'segmentation', or combined into larger units, called 'blocking'. The choice of which of these methods to use will dependent on how to make the most efficient use of the available lower layer service.

When an (N+1)-entity requests a service from an (N)-entity it must utilise the appropriate service-access-point (SAP). SAPs are provided on the layer boundary, and in the case considered here (N)-SAPs (ie those yielding access to N entities) are located at the N and N+1 boundary (see Figure A.4.5). In order

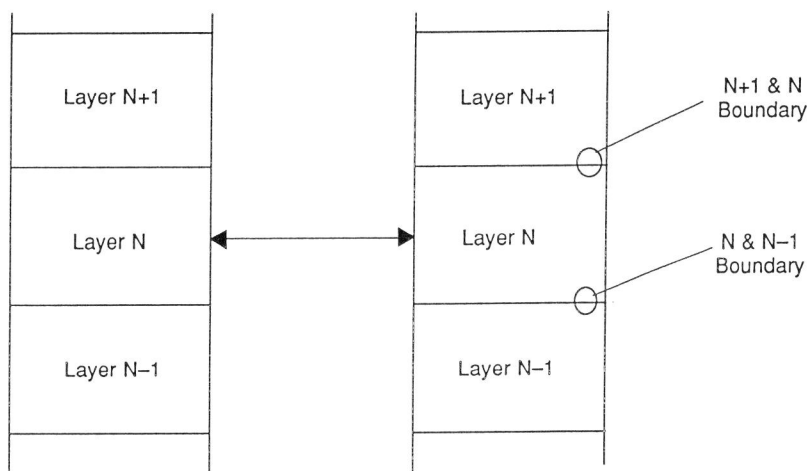

Figure A.4.4   Example of an Arbitrary Layer N Within the Communications Model

to access a specific SAP the (N+1)-entity must utilise the appropriate (N)-address or (N)-service-access-point address, to use its full title.

Description of the Individual Layers Within the ISO Seven-Layer Model

The physical medium in Figure A.4.3 joins the systems involved. It is usually the PTT-supplied network, although it could be a private network such as a local area system. Using CCITT terminology, the boundary is usually the boundary between the data circuit-terminating equipment (DCE)—the modem for example —and the data terminal equipment (DTE)—ie the subscriber's terminal or computer. The functions of each of the individual layers are as follows:

Physical Layer

The physical layer contains functions required for the translation of information or data stream into the form that will achieve transmission, eg electrical impulses, modulation of a carrier wave, etc. This layer provides the electrical and mechanical means for carrying out communications; it is also concerned with procedural aspects of establishing a connection. The service provided by the physical layer to the layer above is the most basic, that is, a serial data stream.

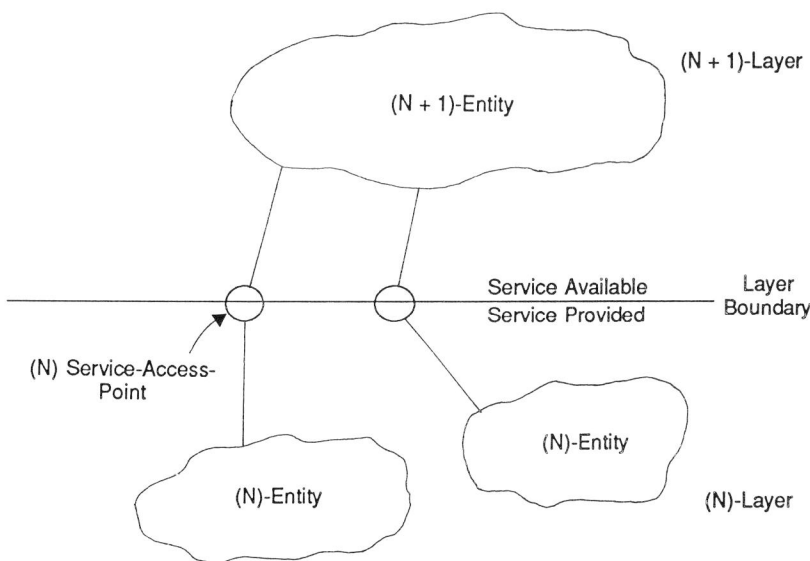

Figure A.4.5   SAPs at a Layer Boundary

Many of the standards in this area pre-date the OSI model and have generally been developed by the CCITT. At this level the standards refer to both modem modulation (eg V.21 and V.27 modem standards), and specifications for interfaces (eg X.21 and X.21 bis, V.24, etc). Currently, work is underway on the development of ISDN (Integrated Services Digital Network) interface standards (CCITT I-series of recommendations) because the emergence of these networks is likely to have some impact on the future of Wide Area Networks (WAN).

(*Note:* V.24 lists the handshakes between Data Terminal Equipment-DTEs, X.21 defines the interface for accessing public data networks, X.21 bis defines the interface between public data networks and V-series modems.)

Data Link Layer

The principal function of the data link layer is to ensure that data passed to the transmission medium achieve a reliable transfer across the network and arrive in an uncorrupted form. Because of the multiplicity of types of linking networks that can be utilised during a transmission it is essential that this function is achieved regardless of the medium being used. The reference model identifies numerous functions for this layer, but two of them in particular are especially important:

— framing – this identifies individual messages and groups within a data stream;

— error detection – physical channels are generally unreliable hence a means of identifying errors is essential.

The recognition that physical transmission mediums are subject to errors, caused by noise and transients produced by inductive means or otherwise, implies that error detection and hence recovery is required. Data link protocols were introduced some time ago to enable errors to be detected and corrected, usually by requesting retransmission. High-level data link control (HDLC) is the most popular example which uses a cyclic redundancy check (CRC) to identify error conditions. CRC employs a mathematical function to generate an error detection code to be passed with the data stream. Its advantages lie in its extremely high percentage error detection rate and the fact that it introduces relatively little redundancy (bits not conveying information) to the data stream and hence improving efficiency.

Network Layer

The function of this layer is to create, maintain and eventually terminate a logical path through a network which can be utilised by the transport layer. In particular it performs addressing, switching, routeing and facility selection functions associated with establishing and operating a connection between systems. The intention is to provide a service to the transport layer independent of the subnetwork(s) which have been used to transmit the data. Included in the functions undertaken by this layer are:

Routeing – decides how to transmit a frame between source and destination using network addresses;

Relaying – enables data transfer across intermediate subnetworks (see Figures A.4.6 and A.4.7);

Flow control – the ability to match traffic flow with the capacity of the transmission path;

Sequencing – ordered delivery of data across the network connection.

The network layer is relatively complex in as much that it has its own internal structure, comprising a number of sublayers. Many of the complications are introduced by the differing requirements of connection-oriented and connectionless services. There are a number of draft proposals in existence from ISO in this area.

The CCITT recommendation X.25 is an interface standard for connection to the public packet switched data networks. This standard covers layers one to three in the OSI model, with the packet level (layer 3) being an example of a possible network layer standard. One of the basic aims of the packet level is to multiplex a number of logical connections over one physical link. At the physical layer, X.25 will make use of X.21, which defines an interface to operate synchronous mode communications over a Public Data Network (PDN). At its second level, the data link layer, X.25 uses a subset of HDLC.

In the same way that X.25 recommendation covers the first three layers of the model the work on ISDN interface (CCITT I-series) and the local area network (LAN) standards (IEEE 802-series and ISO 8802-series) will also cover these layers.

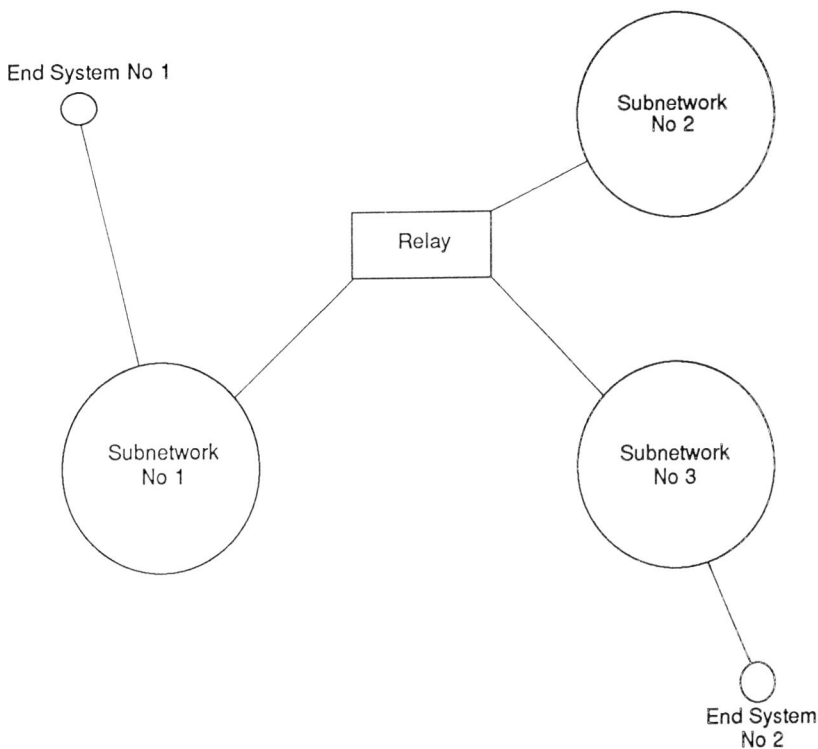

Figure A.4.6   The Requirement for Relaying and Routeing

Transport Layer

The main function of this layer is to create a link for the transport of data between the communicating end systems. It accomplishes this by using the physical transport mechanisms made available by the network. The transport layer is the lowest which exists in end systems only (refer to Figure A.4.7) and therefore performs functions which are of a truely end-to-end nature.

The transport layer provides, in association with the layers below it, a universal data transport service

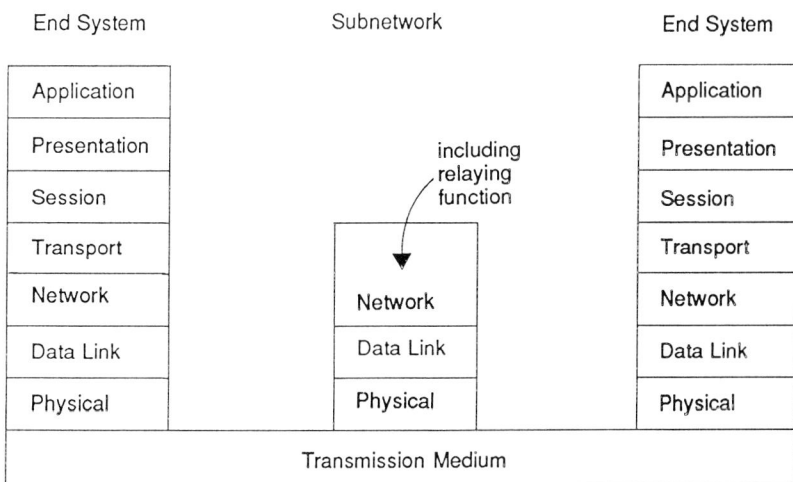

Figure A.4.7   Logical View of an Intermediate Subnetwork

which is independent of the physical medium actually in use. Layers above this service will request a particular class and quality of service, and the transport layer is responsible for optimising the available resources to provide the service requested. The quality of a service is concerned with data transfer rate, residual errors allowed, and associated items, whereas the class of service covers the different types of traffic which diverse applications require—eg batch or transaction processing. The services provided fall into one of five categories numbered 0 to 4. The facilities provided by the lower class numbers are in fact subsets of those provided by the higher classes, with class 0 merely being the service provided by the network layer as this adds no functionality to the lower layer. In contrast to class 0, class 4 employs a powerful protocol to ensure that if any loss, re-order or duplication of data occurs it can be dealt with so that the application will be provided with uncorrupted data. Briefly, the functions provided by each class level are as follows:

— class 0: basic network service;

— class 1: class 0 + error detection and recovery;

— class 2: class 0 + multiplexing;

— class 3: class 1 + class 2, ie error detection, recovery and multiplexing;

— class 4: detection of errors not signalled by the lower layers, correction, recovery and multiplexing.

Session Layer

This layer is responsible for establishing and maintaining a relationship between the end-users wishing to exchange information. Two applications must have a formal liaison for this purpose, which is called a session. The session layer establishes, breaks and maintains this liaison, and ensures that data reaching a system is routed to the correct application. It also ensures that the information exchanged is correctly synchronised and delimited so that, for example, two applications do not try to transmit simultaneously.

The session layer is the lowest, in a communicating stack, which deals explicitly with the communication process between open systems rather than just the conceptual interconnection of such systems. The service it provides to the presentation layer is the management of the dialogue between communicating end systems. The functions performed at this layer depend largely on the specific requirements of the application which is in operation. From the range of services that can be provided the following are significant:

— establishment and close down of the connection;

— synchronisation of the data communication process to allow for error checking and recovery;

— negotiation of the manner of interaction between presentation entities, which can be simultaneous both-way transmission or two-way alternate direction transmission (ie full and half duplex respectively).

Presentation Layer

The applications involved in data communication, in an OSI environment, should not be aware of the characteristics which are not relevant to the actual data being exchanged. If this can be achieved it allows the language of the communication to be transparent to the applications which are involved. The presentation layer performs the two-way function of taking information from applications and converting it into a form suitable for common understanding (ie not machine-dependent), and also presenting data received to the applications in a form they can understand. The layer provides services which give independence from internal character formats (ie transparency), and consequently it provides machine independence.

When a communication link is established, the end-systems will negotiate the encoding rules which are to be utilised during the information exchange. At this stage, any conversions which need to be undertaken will be agreed and carried out at this layer.

Two technically aligned standards have evolved in this area. CCITT recommendation X.409 first and then ISO adopted the words and principles for their two-part standard entitled *Abstract Syntax Notation.1*

(ASN.1). This employs the concept of dual levels of syntax – abstract and transfer. At the abstract level data or information is described in terms of its logical structure and its type. When the abstract syntax has been produced it can then be encoded into a data stream which is the transfer syntax. The relationship which the presentation layer establishes between the abstract syntax and the transfer syntax is called a 'presentation context'. Within a connection there can be multiple presentation contexts which will account for the situation when a number of different abstract and transfer syntaxes are developed and used together.

Application Layer

The Application layer differs from all the rest in that it is the highest layer of the model and therefore does not provide a service to a layer above. It is the source of all data which is to be transported in the OSI environment and ultimately its destination. All the other layers in the model exist to support this layer which in turn exists to provide the information exchange functions between user application processes. It is here that a decision is made whether to treat the communication as a file transfer, virtual terminal session or an MHS interpersonal message.

Because of the requirement for the multiple specific applications to reside at this layer, a formal structure was required. This would specify exactly how application associations (ie communications between applications in separate end-systems) should be formed using particular elements. Unfortunately this area has been the source of some confusion mainly because the original modelling concepts which were proposed have sinced been changed in order to encompass new ideas.

Initial ideas surrounded the use of two sets of basic elements with which associations could be formed. The first set of elements relate to the establishment and release of connections forming a basic 'tool kit' called Common Application Service Elements (CASEs). The second set is specific to certain jobs or activities and are called Specific Application Service Elements (SASEs).

CASE

CASE functions, as presently defined, are further subdivided into two categories:

1 Association Control

These services allow an application to 'Begin', 'End' and 'Halt' Presentation connections.

2 Commitment, Concurrency and Recovery (CCR)
These services allow an application to have control over the 'atomic actions' of a connection.

SASEs

SASE functions cover specific jobs, tasks or activities. At present ISO work in this area has been on bulk file transfer (known as FTAM), job execution upon remote systems (JTM) and the use of a terminal type which is application-independent (VT). It is likely that other ISO developed SASEs will be advanced to cover other tasks and it is also likely that the private sector will develop their own SASEs for industry-specific applications.

In the context of Message Handling Systems based on the 1984 X.400 recommendations, they do not fit into the CASE-SASE applications model. Instead the concepts of the Remote Operations and Reliable Transfer Service (RTS) are employed to give the application entities, SDEs and MTAs, a simplified interface to the session layer with a minimal presentation layer service.

This original model has now been changed and is re-specified as the generic application layer structure within a new standard, ISO/DIS 9545. In the OSI environment communication between application processes is now represented in terms of an association between Application Entities (AEs) residing within the application layer using a presentation layer connection. A user's local process will interact with the application layer via a User Element (UE) which has an AE associated with it, see Figure A.4.8 . The interaction between AEs is described in their use of Application Service Elements (ASEs) which are sub-divisions of the total functionality of the whole AE. An ASE is a grouping of functionality which will support a typical application. Each ASE represents a different kind of work which the user requires to be performed. ASEs come in basically two types, these are:

Re-usable ASEs. – These are analogous to the concept of CASE functional elements, as specified in the early model, because they provide generic services which are 're-usable' in many types of association. Examples of these would be ACSE, CCR, ROSE and RTSE.

Application Specific ASE. – These are analogous to the SASE functional elements, as specified in the early model, because they are specific to a particular job or function. The application-specific ASEs may have their functionality divided still further into other ASEs which deal with specific aspects of a service to be provided. Examples of these would be FTAM, JTM, VT and MHS. The MHS ASE, from the 1988 recommendations, has been sub-divided into other ASEs, such as ASEs used for access to the Message Transfer (MT) service:

- Message Submission Service Element (MSSE);

- Message Delivery Service Element (MDSE);

- Message Administration Service Element (MASE).

The combination of ASEs used in a specific communication scenario is termed an application-context. The choice between which application-context is to be used will be dependent on the type of work that the application process has requested to be completed via the UE.

Although the situation with two successive models for the application layer structure appears unnecessarily complex, it is not a source of potential worry. It should be emphasised that a change in the modelling concepts does not definitely imply changes in real implementations. The model is only a means of conveying the concepts behind a more complex operation which, in this case, remains unchanged although the model has been updated.

The following sections discuss the ISO developed ASEs which are now available.

File Transfer Access and Management (FTAM)

FTAM is the ISO OSI standard to allow open systems within a distributed computer system to transfer, access and manage files. Some examples of the types of 'real time' applications which might employ this standard are:

— complete file or bulk file transmission;

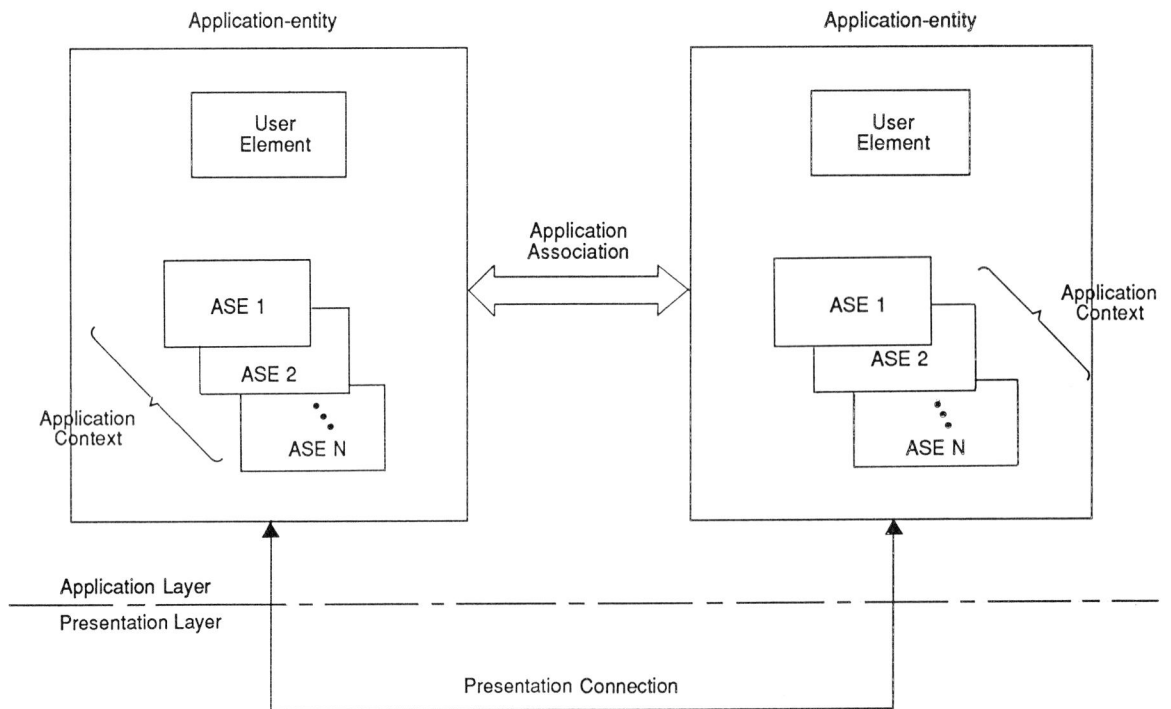

Figure A.4.8   The Application Layer Model

Appendix 5
Functional Standards or Profiles

In order that OSI standards will be applicable to both systems today and those that will be developed in the future they have been designed so that they have a wide range of options which allow them to cope with many conceivable situations. Because these base documents have little or no pragmatic constraints it is left to the implementors to introduce sensible constraints and select options as they require. However, if each individual implementor was only to add constraints and choose the options which they and they alone require, it would quickly lead to a situation where supposedly open systems were incompatible.

To avoid this potentially troublesome situation, various organisations have proposed functional standards or profiles which are derived from the base documents. In this way, systems based on the same profile will be compatible.

PURPOSE OF A PROFILE

The purpose of a profile is effectively to take a slice through the base standards so as to make them applicable to real world situations. A profile may be of one individual base standard, a stack of all seven layers, or some number in between these two. The first stage in the development of a profile is to choose which standards are to be used for the individual layers: this forms a 'protocol stack'. This can then be further refined by the addition of pragmatic constraints (eg setting the maximum permissible length of message in a messaging system) and choosing which options are to be included from the base standard/s. When this refinement process is complete the resulting document is a profile.

The major uses for the profile are in the procurement and conformance testing of systems. In procurement a purchaser can demand that all their systems should conform to a particular specification in order that compatibility is maintained. Typically, manufacturers claim that their OSI products conform to one or more specific profiles, so it is possible to see at a glance whether a product is applicable or not. Because a profile, like any other standard, is produced in a human language it is open to ambiguities and misinterpretation. This situation will exist until standards are defined by mathematically based formal methods which prevent such problems. For the moment, however, the only way to prove conformance to a profile or base standard is to conformance test the implementation. Conformance testing (see Appendix 6) will ensure that a product conforms to a specific profile.

The major problem with the creation of profiles and functional standards has been the complete lack of meaningful co-ordination. Many organisations are profiling the same base documents and producing work which is incompatible at a detailed level. In the X400/MOTIS messaging arena alone CEN/CENELEC, CEPT, NIST, British Telecom (ONA) and UK GOSIP have all produced profiles of the base documents, the main differences here being between the European and NIST North American work. There is an obvious requirement for co-ordination and this now seems to be forthcoming from ISO JTC1 who have set up a Special Group on Function Standardisation. This group was created specifically to deal with the taxonomy (listing and classification) of functional standards. Eventually it is intended that one profile will be registered for each specific application area and these will be known as Internationally Standardised Profiles, or ISPs.

Figure A.5.1 shows a map of Europe and the United States of America indicating the groups working on functional standardisation topics. The diagram indicates the relationship between these developments.

EUROPEAN FUNCTIONAL PROFILES

The concept of functional standards evolved from the Standards Promotion and Application Group (SPAG). This is a body which has been formed by a group of 12 European suppliers. Additionally it is supported by the participation of European standards groups such as the BSI and is subject to their review and approval. More recently its membership has expanded to encompass non-European manufacturers.

Other examples of the work within Europe on functional standards are from groups such as CEN/CENELEC and CEPT in the area of European Norm's (EN's) which are meant to be the profiles which all European organisations will use in the procurement of OSI products. This point should be of particular interest to local authorities after the recent EEC decision that their future IT purchases must acknowledge the use of OSI where practical. Another group to be considered is ARPOSE, in France, who are working on OSI profiles up to the transport layer and making contributions to the US NIST Workshops and their functional standards.

British Telecom have also carried out work on functional standards, producing their Open Network Architectures (ONA) profiles. A recent application of this has been their message handling system, Gold 400.

US FUNCTIONAL PROFILES

The most significant activity from the United States in the area of functional standardisation has been from General Motors (GM) and The Boeing Company with their Manufacturers Automation Protocol (MAP) and Technical Office Protocols (TOP), respectively.

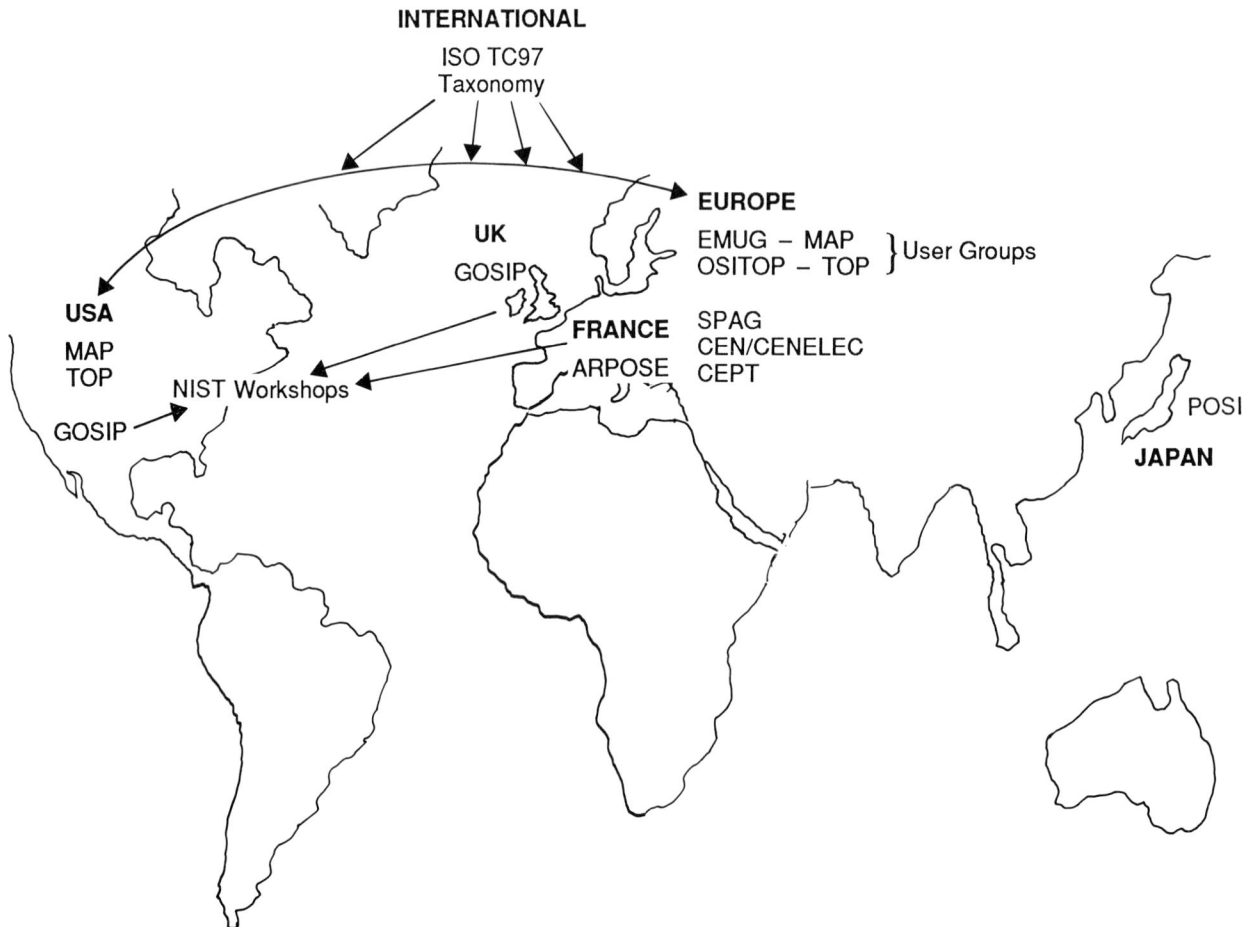

Figure A.5.1   Worldwide Functional Standards Activity

In 1980 GM commenced work on MAP in response to problems of interconnection between their advanced plant-floor manufacturing equipment. MAP draws from both OSI and proprietary protocols to define a specification suitable for the procurement and testing of equipment. As OSI standards become more stable, future versions of MAP will be enhanced to adopt these new recommendations.

The TOP program, organised and administered by The Boeing Company, is open to other members such as users and vendors. The aim of the program is to produce a profile of the protocols for the communication of applications in a technical office environment.

A measure of the interest in the MAP and TOP work can be gained from the powerful user groups which have been formed both in the United States and in Europe. The US MAP/TOP group (joint MAP/TOP Steering Committee) and in Europe the European MAP User Group (EMUG) and OSITOP (European TOP user group) have been formed so as to allow the users of systems based on these profiles a forum for discussion and focal points for feedback to the groups working on the profiles.

NIST are also carrying out functional standardisation work, notably on the CCITT's X.400 recommendations, in their Workshop forums.

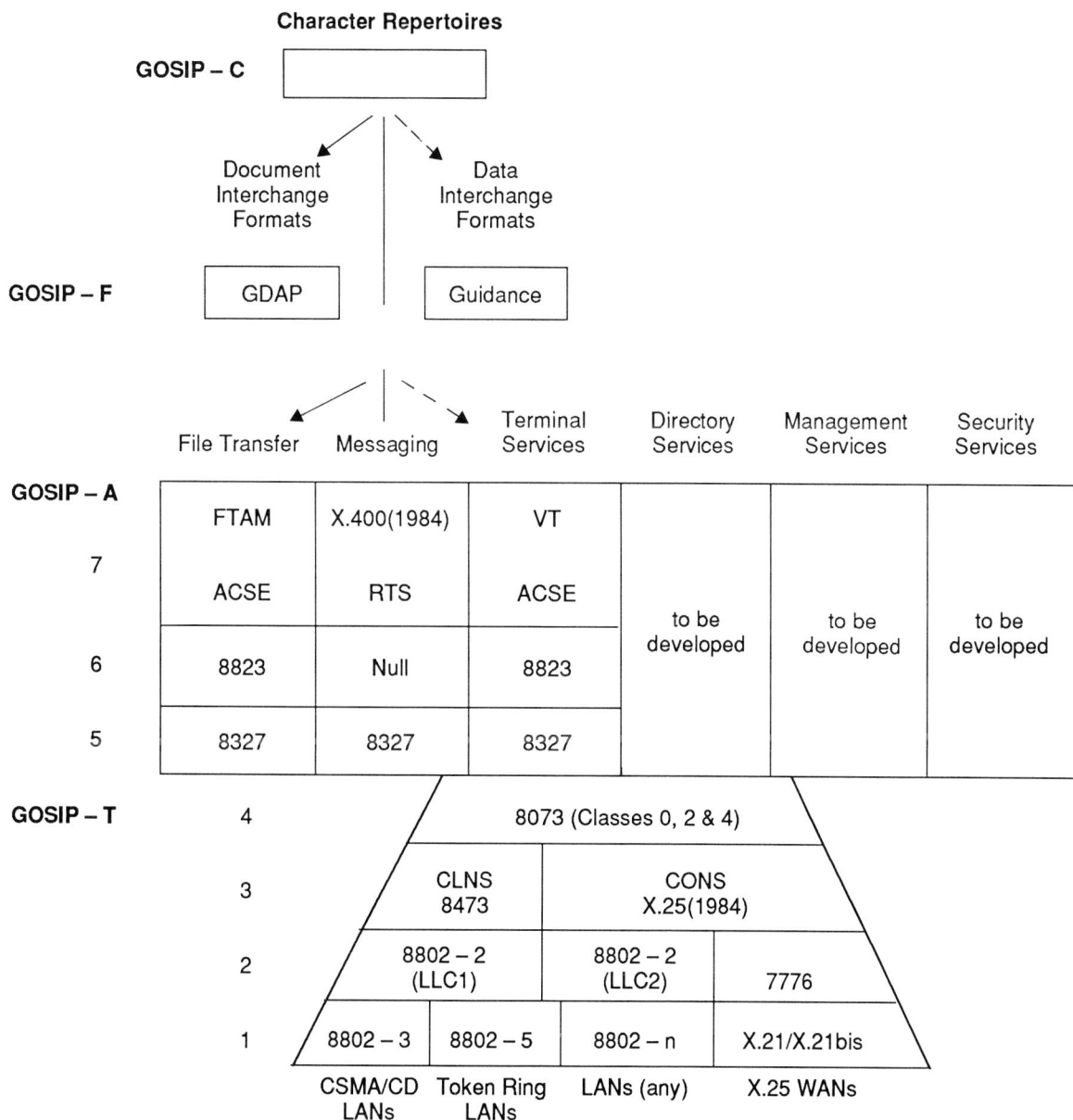

Figure A.5.2   The UK GOSIP Architectural Model

Other US interest in profiles has been focused through the Corporation for Open Systems (COS) established in North America to advance the movement towards OSI and in particular X.400 message handling. COS are effectively SPAG's counterpart with its members being drawn from leading US vendors, service providers, Universities and users.

WORLDWIDE GOVERNMENTAL INTEREST

In the United Kingdom the CCTA have attempted to produce a profile for government departments seeking to purchase OSI products, again particularly relevant to local authorities since the EEC decision on the IT purchases of government departments. The profile is called the Government OSI Profile, or GOSIP for short. GOSIP has three defined aims which are:

— facilitate procurement and acceptance testing;

— ensure that separately procured systems are capable of interworking;

— provide direction to manufacturers on product development.

Figure A.5.2 shows the development of the GOSIP Architectural Model to date. The US also have a counterpart to the GOSIP project which is called US GOSIP. This has similar aims to the UK work.

In Japan the Ministry of Posts and Telecommunications have produced their own OSI profile of layers 1 to 5, specifically for personal computers. The Japanese Unified Standards for Telecommunications - Personal Computers, or JUST-PC for short, acts as the basis for their JUST-Message Handling System (JUST-MHS) work which profiles the CCITT X.400 recommendations.

Appendix 6

Conformance and Interoperability Testing

For two product vendors to claim conformance of their products to a specific functional profile is not in itself sufficient to ensure that they will interwork successfully. Even though profiles effectively 'slice through' the protocol stacks, pairing down the options, they are still produced in a normal language and so are open to human interpretation. Advances in the use of mathematical formal description languages within standards will almost certainly help here, but for the present they are some way off. The only way to ensure that implementations will interwork successfully is to test their conformance to the profile and then to carry out interoperability testing.

Conformance testing can be subdivided into two categories:

Static Conformance Testing. This is effectively a paper exercise whereby the functions provided by the standardised protocol are checked against those claimed by the product. These claims are detailed in a Protocol Implementation Conformance Statement, or PICS.

Dynamic Conformance Testing. This tests the behaviour exhibited by an Implementation Under Test (IUT). The IUT is treated as a 'black-box' and its visible inputs and outputs are monitored by a test suite.

As an aid to OSI conformance testing ISO is currently developing a standardised framework and methodology. This standard when complete will be published in six parts, currently DP9646/1—6. This methodology work has formed the basis of the CCITT's recently finalised draft X.403, the Conformance Testing of 1984 X.400 systems.

Because of the difficulty in simulating all practical operating conditions during conformance testing, it is necessary to resort to another level of testing to ensure effective interworking. Interoperability testing allows the two systems which are to interwork to be connected and exercised in a controlled manner by the tester which will control and monitor the link between them.

Only when conformance and interoperability test centres have been successfully established will the truly open system become a reality. It is encouraging to see that the seeds have already been sown in Europe and in the US with the establishment of national and private test centres. These will be discussed in the next sections. Interworking demonstrations which have been taking place at exhibitions, such as Autofact 85 (MAP0, ceBit 1987 (14 X.400 participants) and Telecom 87 (21 X.400 participants) and many others, are all useful dry runs which will start to iron-out product incompatibilities.

WORLDWIDE CONFORMANCE TESTING ACTIVITIES

The National Physical Laboratory (NPL), in the UK, began research into OSI conformance testing as early as 1979. The NPL is the industrial research establishment of the Department of Trade and Industry (DTI) which meant that it was outside its remit to offer a commercial OSI testing service. Thus in 1981 the NCC was selected as an organisation to progress the work towards the provision of an OSI Transport Testing Service. In 1984 this was achieved and entitled NCC COMMS-AID.

Because of the likely volume of testing work to be completed in the future, the DTI decided that another centre would be required in the UK. The Networking Centre established in 1985, with DTI and industry funding, to undertake OSI LAN testing, ie based on the IEEE 802 series. Since then the centre has been contracted by the European Commission, under its Conformance Testing Services—Local Area Network (CTS—LAN) initiative, to be a European test centre for LANs.

In 1986 the European Commission also placed contracts for the establishment of European harmonised Conformance Testing Services for Wide Area Networks (CTS-WAN) covering the following areas:

— OSI (layers 1 to 4);

— teletex (layers 5 to 7);

— message handling systems (layers 5 to 7);

— file transfer, access and management (layers 5 to 7).

These contracts involve organisations from many different countries, including France, Denmark, Italy, Federal Republic of Germany and the UK, with the NCC and British Telecom.

Standards Promotion and Action Group (SPAG) Services is an offshoot of SPAG and their activities involve interworking tests and services, development of support services, advice/expertise and information provision. SPAG Services X.400 test capability was used in the preparation for the Hanover ceBIT X.400 demonstration, with all of the 14 participants being connected at some stage.

The Swiss PTT, like BT in the UK with X400 Teleprove, are offering an X.400 testing service, called OSI-Lab, in addition to their 'arCom 400' message handling service.

The Corporation for Open Systems, in the United States, is also taking an active role in the provision of testing services and is to develop, with the NCC in the UK, X.400 testing services. COS are also heavily involved with testing for the Enterprise Networking Event in the US (June 1988) where MAP V.3.0 will be demonstrated for the first time.

Appendix 7
X.400 Technical Guide

Chapter 3 of this book introduced the technical concepts of Message Handling Systems (MHSs) based on the CCITT's X.400 recommendations. This chapter will add more detail to this original description by looking at the following areas:

— facilities provided by the MHS services;

— X.400 implementation – the options;

— remote User Agent (UA) Access;

— remote operations;

— reliable transfer server;

— use of OSI lower layers by X.400;

— routeing within an MHS;

— directory functions within an MHS;

— relationship between the major X.400 profiles.

All of these issues will be of interest to organisations looking towards implementing X.400 systems. However, it must be emphasised that it is advisable to obtain a copy of the recommendations and implementors' guides before proceeding with equipment procurement. Careful study of these documents will show all the available options within the standard and will allow a detailed list of requirements to be specified.

THE FACILITIES PROVIDED BY MHS SERVICES

In earlier chapters of this book MHS services were considered at a generic level, looking at their overall purpose rather than the detailed facilities contained therein. This section will list the facilities provided by the Message Transfer (MT) service and the Interpersonal Messaging Service (IPMS).

MT Service

This is the generic service operated between MTAs and UAs to provide an application-independent, store-and-forward message transfer service. The main aim of this service is to move messages consisting of information and/or data between the originator and the recipient. It provides the UAs with access to the MTS in order to exchange messages.

The MT service consists of a number of service groups which, except for the 'Basic' group, are optional. These groups are in turn subdivided into a series of service elements, each of which performs a specific function. Each service group is made up of elements which together perform a specific aspect of the MT service. The service groups which have been defined are:

Basic. These are concerned with the unique identification of the message, the date and time of message delivery, and submission and the type of contents.

Submission and Delivery. These deal with the specification of the type of delivery required (eg speed, number of recipients, etc.) and whether delivery notification is applicable.

Conversion. Whether or not message content conversion is required, for example in the case of delivery to a telex terminal where ASCII text would need to be converted to the telex character set.

Query. To establish if message delivery is possible, ie is the message suitable to be received by the recipient's terminal equipment.

Status and Inform. This group allows a message with certain attributes to be delivered only to specific UAs. Also gives a UA the capability to indicate to the MTS that it is not ready to accept message delivery.

After looking at the general groups within the MT service the following is a complete list of the features provided by this service. A full explanation of all of the following is given in recommendation X.400 sections 4.1.1 to 4.1.5.2:

Service group	Service elements
Basic	Access management
	Content type indication
	Converted indication
	Delivery time stamp indication
	Message identification
	Non-delivery notification
	Original encoded information types indication
	Registered encoded information types
	Submission time stamp indication
Submission and delivery	Alternate recipient allowed
	Deferred delivery
	Deferred delivery cancellation
	Delivery notification
	Disclosure of other recipients
	Grade of delivery selection
	Multi-destination delivery
	Prevention of non-delivery notification
	Return of content
Conversion	Conversion prohibition
	Explicit conversion
	Implicit conversion
Query	Probe
Status and inform	Alternate recipient assignment
	hold for delivery

Interpersonal Messaging Service

The Interpersonal Messaging or IPM service provides its users with electronic mail/messaging facilities which bear some resemblance to existing public and private electronic mail systems. The service is built upon the MT service and is provided by a class of UAs called IPM UAs.

IPM UAs make use of the 'Basic' group of MT facilities while also permitting the optional ones to be requested. In addition to these service elements the IPM UAs also provide other capabilities which make up the IPM service. These additional capabilities are again separated into groups as follows:

Co-operating IPM UA Action. These are service elements which involve the IPM UA in some form of action upon the message, eg receipt notification, auto-forward indication, etc.

Co-operating IPM UA Information Conveying. These service elements are concerned with the identification of the originator/recipient and status of a message.

The elements of the IPM service are:

Service group	Service elements
Basic	Basic MT service elements
	IP message identification
	Types body
Submission and delivery and conversion	As per MT service
Co-operating IPM Action	Blind copy recipient indication
	Non-receipt notification
	Receipt notification
	Auto-forward indication
Co-operating IPM UA Information Conveying	Originator indication
	Authorising user's indication
	Primary and copy recipient's indication
	Expiry date indication
	Cross-referencing indication
	Importance indication
	Obsoleting indication
	Sensitivity indication
	Subject indication
	Replying IP message indication
	Reply request indication
	Forwarded IP message indication
	Body part encryption indication
	Multi-part body
Query	As per MT service table
Status and inform	As per MT service table

X.400 IMPLEMENTATION OPTIONS FOR PHYSICAL SYSTEMS

The message handling model introduces the concept of sublayers within the OSI application layer (see Figure A.7.1). The first of these layers is the User Agent Layer (UAL) within which the User Agent (UAs) reside. The second layer is the Message Transfer Layer (MTL) within which the Message Transfer Agents (MTAs) and Submission and Delivery Entities (SDEs) reside. The reasons for the introduction of these layers are as follows:

— allows a boundary to separate groups of similar functionality;

— minimises the interactions across layer boundaries;

— the separation allows different protocols within layers to be used without affecting each other.

Because of the separation caused by sublayering the OSI application layer, the concept of functional entities can be introduced. These are the MTAs, SDEs and UAs, which are merely groupings of similar functionality to be provided by a system. And because these particular entities reside in the application layer they are normally referred to under the generic title of Application Entities—or AES.

The separation between the MHS functional blocks allows various option when implementing a real system. The UA and MTA may either be co-located (ie resident within the same system) or implemented on separate hardware. If co-located, the interface between the UA and MTA is left for the implementor to define as it will have no externally detectable effect on their operation. If the MTA and UA are to be standalone items then their interface has to be standardised in order to allow equipment from different vendors to be used, eg an MTA from vendor A and a UA from vendor B. To achieve this, a Submission and Delivery Entity (SDE) and its associated P3 submission and delivery protocol are introduced. The SDE is resident with the remote UA and allows it access to the services of the MTL available at the UA-MTA boundary. The SDE does not provide message transfer services itself but rather interacts with the MTA to access them remotely.

The mapping of the elements of the MHS on to physical systems can be accomplished in a variety of ways and these are indicated in Figure A.7.2. It can be seen that it is possible to have any of the following options:

— UA and MTA co-located on the same hardware with the user being some device which can interact and perform I/O with the UA;

— UA and MTA not co-located. In this case the UA and associated SDE would be implemented on some other intelligent hardware such as a personal computer or minicomputer. In the case of the latter it may have several remote UAs implemented on it.

REMOTE USER AGENT ACCESS

Provisions are made, within the model, for the option of remote UA operation; remote meaning that the MTA and UA are not co-located. In such cases the remote or standalone UA is split into two functional entities:

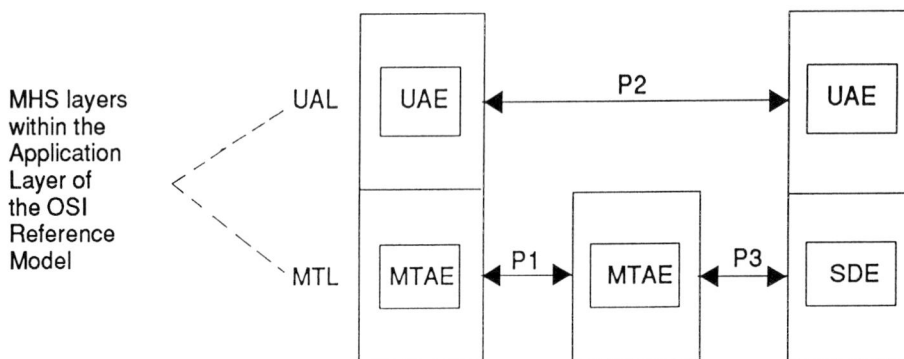

Figure A.7.1   MHS Sublayering Within the OSI Application Layer

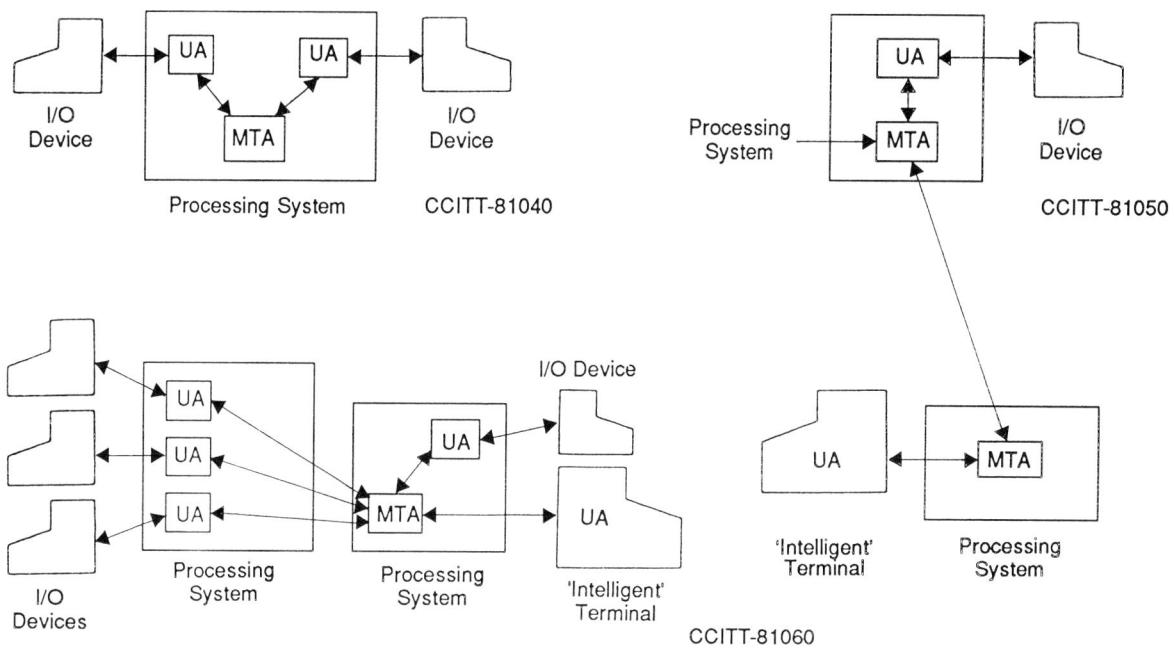

Figure A.7.2   Implementation Options (Source X.400)

— the User Agent Entity (UAE);

— the Submission and Delivery Entity (SDE).

The UAE provides the user with the normal UA facilities and functions, while the SDE provides the UAE with access to the facilities of the Message Transfer (MT) service. Whereas message transfer between MTAs is based on the store and forward principle SDE and MTA interaction is interactive in the same way that it would be when the MTA and UA are co-located.

One of the most glaring omissions from the original 1984 X.400 recommendations is the lack of a message store buffer which can be used when a UA is remote from its MTA. When a message is received by the recipient's MTA it is always delivered directly to the UA or remote UA. This mode of operation is perfectly acceptable when UA and MTA are co-located because when one element is active the other is as well. Hence when the UA receives a message it will store it in its own message store until the user becomes active and accesses it. In the case of the remote UA this instant delivery of a message from the MTA to the UA assumes that it is always attached and active. This is a fair assumption if the remote UA is implemented on a minicomputer, for example, which is active 24 hours a day. But if it is implemented on a microcomputer it is not likely that it will remain active 24 hours a day or even every day (eg if the human user is on holiday). In this situation the messages will be held at the MTA until such time as the remote UA becomes 'live'. At this point the MTA will attempt to deliver all the messages for the user to his UA. This situation is still acceptable unless the UA happens to have been off-line for some time. The MTA will attempt to deliver a large number of messages and flood the storage area. If there is insufficient storage area available some messages will be lost. This serious problem can be overcome if the user has a personal message store buffer resident at the MYA. This would allow the user to selectively receive messages in an orderly manner rather than being overwhelmed.

This approach reduces the potential usefulness of the remote UA approach, a fact which was quickly realised, particularly as the move towards ever more complex microcomputers continues. The solution to this situation is provided by the 1988 version of the X.400 recommendations which include message store concepts (see Chapter 10).

REMOTE OPERATIONS

When a UA and MTA are co-located the communication process between them is interactive, ie UA submitting messages and MTA delivering. In order that the same situation will apply when a remote UA is communicating with its MTA the interactive P3 protocol has been designed. P3 allows the remote UA to access the MTL facilities from its MTA. This access includes the transfer, between the SDE and MTA, of messages and responsibility for them during both the submission and delivery phases.

Remote operations as defined within recommendations X.410 provide a framework for the specification and implementation of interactive application layer protocols such as P3. It specifies a form of remote procedure call mechanism which will allow an operation to be invoked, along with any required arguments. Remote operations also make provision for the results of an operation returned after the completion. In this way the SDE communicates with an MTA, during the submission phase, by remotely invoking operations within it, while during the delivery phase the MTA will invoke operations within the SDE.

When any Application Entity (AE) invokes an operation which is provided by another it does so by entering into an exchange of Operation Protocol Data Units (OPDUs). Remote operations defines four such OPDUs:

Invoke. This is used to invoke some operation and to pass any parameters associated with the operation.

Return Result. This is returned to the invoker of an operation upon its successful completion along with any results.

Return Error. This signals an unsuccessful termination of an operation and passes back diagnostic data.

Reject. This is a 'catch-all' for responding to an unexpected ot indecipherable message.

These four basic messages are transferred over the reliable transfer server, as defined within X.410, which is considered in the next section.

RELIABLE TRANSFER SERVER

The reliable transfer server, or RTS, is that part of an application entity (AE) which is responsible for creating and maintaining associations between an AE and its peers. The AEs in this X.400 context are of course the MTAs and SDEs. The other parts of an AE, in addition to the RTS, are termed the RTS users and are those which drive the application protocol, in this case either P1 or P3.

An association is created between AEs in order to transfer Application Protocol Data Units (APDUs). RTS will create such an association, maintain it and ensure the reliable transfer of APDUs. There are two forms of association which RTS can create:

Monologue – undirectional communications;

Two-Way Alternate – bidirectional communications but only in one direction at any single instance.

The two-way alternate mode of communicating requires the introduction of a 'turn' system whereby an AE can only communicate with its peer in the association when it is in possession of the 'turn'. The RTS and RTS-user interactions are described as a set of service primitives, these are:

Open – establish an association;

Close – release an association;

Turn Please – request for the exchange of the turn;

Turn Give – exchange of the turn;

Transfer – reliable transfer of an APDU;

Exception – indication of transfer failure.

The *Open* and *Close* primitives are used by an MTA or SDE to establish or terminate, respectively, an association with a peer entity. As indicated earlier, the turn commands act effectively as a flag between the AEs to indicate who can communicate when operating in two-way alternate mode. *Transfer* requests the reliable transfer of data while *Exception* indicates when a transfer of data has failed.

X.400 – USE OF THE LOWER OSI LAYERS

The X.400 recommendations define an application layer protocol to achieve message handling functionality. In order to transfer data and protocol information between the AEs of an MHS the lower layers of the OSI model must be employed. The requirements of layers 1 to 6 placed upon the OSI model are defined within recommendation X.410 sections 4 and 5. These are summarised below:

Presentation Layer – minimal or null presentation protocol defined within section 4.2.1. of X.410 using X.409 notation;

Session Layer – extensive use is made of this layer using a subset of the session layer BAS (Basic Activity Subset) services;

Transport Layer – X.400 utilises Class 0 which provides the basic network services without additional error detection and recovery;

Network, Data Link and Physical Layers – X.400 makes use of the lower layer procedures defined by X.25, ie packet switched environments.

ROUTEING WITHIN AN MHS

An X.400 Originator/Recipient (O/R) address is a descriptive name for a UA which has characteristics to help the MTS locate the UA within the messaging environment. A 'route' in X.400 terminology is the path to be taken by a message through the MTS in order to reach the recipient's UA. The O/R address will provide the MTS with the information required to select a route.

The X.400 recommendations only specify the method of routeing between the originator's and recipient's management domains. It was felt that internal routeing within such a domain is outside the scope of the standardisation process. The routeing method specified is an incremental one whereby each MD along a given route will determine the MTA within the next MD to which the message will be passed. Routeing in between ADMDs is an issue for bilateral agreement between the two routes. With this method no attempt is made to establish a full route for the message, outside the management domains which are to be used, either by the originating MD or subsequent ones along the route.

DIRECTORY FUNCTIONS WITHIN AN MHS

Appendix 4 outlines some of the activities currently being undertaken by CCITT and ISO in the field of OSI directories. This work will be used as the basis for establishing a fully interconnected and integrated worldwide MHS directory in the future. The 1988 X.400 recommendations have now fully integrated the concepts of X.500 directories within the basic MHS model and it is this that will ultimately lead to the use of user-friendly O/R names within an MHS. The 1988 X.400 recommendations, see Chapter 10 will make full use of directory functions by allowing integrated directory access.

Within an MHS there are a number of uses for a directory which need to be considered:

— helping a user with message preparation, ie returning the O/R name of the recipient;

— establishing distribution lists to be used by the user and MTS;

— returning an O/R address when given an O/R name by the MTS, this will be used by the MHS to establish a route to the intended recipient.

When a user produces a message and indicates the intended recipient by their O/R name the MTS will need to refer to the directory in order to establish the corresponding O/R address. When in possession of the address the required route can be established between the originator and recipient through knowledge of the network topology.

In the early days of an X.400 messaging infrastructure it is unlikely that a fully interconnected and integrated directory service will be available. During this period it will be necessary to employ the extensive use of X.400 O/R addresses, with the migration to using the more user friendly O/R names being a gradual process.

THE PROFILES OF 1984 X.400 AND ISO MOTIS

The profiles of the messaging standards are the documents which the user will specify conformance to when procuring products from vendors. Accurate conformance to these profiles is essential if implementations from different vendors are to interwork successfully. The eventual means of checking this conformance will be via the testing services which are now being developed by organisations such as SPAG Services, the Networking Centre and the NCC.

When looking at the implementation of CCITT 1984 X.400 and early ISO MOTIS (extensions for private messaging, ie inter-PRMD) work there are basically three profile documents to be considered. These are as follows:

CEPT A/311 – this profile carries European Pre-Norm. status (ENV 41202) and as such will eventually become the standardised messaging profile (PRMD to ADMD interworking) for the European Community;

CEN/CENELEC A/3211 – this profile also carries European Pre-Norm. status (ENV 41201) but is complementary to ENV 41202 because it addresses PRMD to PRMD interworking;

NIST NBSIR 86-3386-5 – this is a North American profile for X.400 message handling interworking. It specifies the functions and procedures for PRMD to PRMD and PRMD to ADMD connections.

The CEPT and CEN/CENELEC work is complementary but these and the corresponding NIST profile, which encompasses the same areas, are not compatible at a detailed level. These problems are recognised, however, and there are moves, currently underway, to try to align these documents as soon as possible.

Appendix 8
Directory of X.400 Products

The emergence of X.400 products into the marketplace has now become very rapid indeed, and almost on a monthly basis companies seem to be making new product announcements. Because of this, it is a fruitless task to attempt to give an up-to-date list in a book of this kind. Rather, the approach here will be to give a brief summary of what is available or at least announced at the time of writing, and in this way yield an insight into the rapid adoption of X.400.

When moving towards implementation however, it is essential to check directly with suppliers in order to gain a clear idea of the current range of X.400 offerings.

SUPPLIERS COVERED

The information which has been obtained represents the X.400 products supplied or planned, as at May 1988, for the following suppliers:

British Telecom	ITL
CAP	Logica
Concurrent	McDonnell Douglas
Data General	NCR
DEC	Norsk Data
GPT	Prime
Hewlett Packard	Syndey
Honeywell Bull	Tandem
IBM	Unisys
ICL	Wang

NCC OSI PRODUCT SURVEY

The information contained in this section is an extraction of that obtained during a general NCC OSI product survey completed in May 1988. The sections indicated in this book are those concerned only with functions used to support the X.400 message handling. In addition two important information types, Office Document Architecture (ODA) and Electronic Data Interchange (EDI), are also listed.

SUPPLIER: BRITISH TELECOM

	Product Availability		
	System PC LAN-based X.400 product	*System* VAX-based X.400 product	*System* UNIX-based X.400 product
Architecture	- avail. end 88	ONA avail. end 88	ONA avail. end 88
WANs			
X.25 (1980)	-	-	-
X.25 (1984)	yes	yes	yes
LANs			
8802/3 -	yes	yes	yes
8802/4 -	yes	-	-
8802/5 -	yes	-	-
LLC1	yes	yes	yes
LLC2	-	-	-
N/W Service			
CONS LANs	-	-	-
WANs	yes	yes	yes
CLNS LANs	yes	yes	yes
WANs	-	-	-
Transport Layer			
0	yes	yes	yes
1	-	-	-
2	-	-	-
3	-	-	-
4	yes	yes	yes
Session layer	yes (BAS)	yes	yes
Presentation layer	yes X.410 compatible subset	yes X.410 compatible subset	yes X.410 compatible subset
Application layer:			
File Transfer Access and Management (FTAM)	-	-	-
Job Transfer and Manipulation (JTM)	-	-	-
Office Document Architecture (ODA)	-	-	-
Electronic Data Interchange (EDI)	-	-	-

SUPPLIER: CAP INDUSTRIES

	Product Availability	Comments on Functionality
	System RETIX	
Architecture	Portable C	Portable operating system independent, protocol software
WANs X.25 (1980)	yes	Both HDLC/LAPB and PLP. 1980/84 selectable by date switch
X.25 (1984)	yes	
LANs 8802/3 -	-	
8802/4 -	-	
8802/5 -	-	
LLC1	yes	Independent of LAN media type
LLC2	yes	May also be used with X.25 PLP
N/W Service		
CONS LANs	yes	Over PLP/LLC2
WANs	yes	Over PLP/LAPB
CLNS LANs	yes	Both full and null over LLCI
WANs	yes	Both full and null over PLP/LAPB
Transport layer		
0	yes	Both Connection-mode and Connectionless-mode Transport Service over CONS
1	-	
2	yes	Both Connection-mode and Connectionless-mode over CONS (negotiate down to 0)
3	-	
4	yes	Both Connection-mode and Connectionless-mode (Transport Service) over CONS (negotiate down to 2 or 0) or CNLS (4 only)
Session layer	yes	Sets of functional units for kernel (eg MAP/TOP 3.0) For X.400 MHS or for Reliable FTAM or all functional units
Presentation layer	yes	Kernel (eg MAP/TOP 3.0) + interface for user supplied extensions
Message Handling (MHS)	yes	NBS, CEN/CENELEC and CEPT profiles P1, P2 PRMD-PRMD PRMD-ADMD 1984 (1988 planned)
Office Document Architecture (ODA)	-	
Electronic Data Interchange (EDI)	-	
Other ACSE	yes	
ROSE	yes	
Network Management	yes	MAP/TOP 3.0
Directory Service	yes	MAP/TOP 3.0 (based on X.500)

SUPPLIER: CONCURRENT COMPUTER CORPORATION

	Product Availability
	System OS/32
Architecture	PENnet
WANs	
X25 (1980)	yes
X25 (1984)	planned 89
LANs	
8802/3 -	yes
8802/4 -	planned
8802/5 -	-
LLC1	planned 88
LLC2	yes
N/W Service	
CONs LANs	yes
WANs	yes
CLNS LANs	planned 88
WANs	planned 89
Transport layer	
0	planned 89
1	-
2	planned 89
3	yes
4	yes
Session layer	planned 89
Presentation layer	planned 89
Message Handling MHS X.400	planned 89
Office Document Architecture (ODA)	—
Electronic Data Interchange (EDI)	—

SUPPLIER: DATA GENERAL

	Product Availability
	System DG/X.400
Architecture	
WANs	
X25 (1980)	yes
X25 (1984)	planned
LANs	
8802/3	—
8802/4	planned
8802/5	—
LLC1	—
LLC2	—
N/W Service	
CONS	—
	—
CLNS	—
	—
Transport layer	
0	yes
1	—
2	—
3	—
4	planned
Session layer	yes
Presentation layer	yes
Office Document Architecture (ODA)	planned
Electronic Data Interchange (EDI)	—

SUPPLIER: DIGITAL EQUIPMENT CO

	Product Availability
	System VAX/VMS
Architecture	DNA
WANs	
X.25 (1980)	yes
X.25 (1984)	yes
LANs	
8802/3 —	yes
8802/4 —	yes
8802/5 —	—
LLC1	yes
LLC2	yes (UK only)
N/W Service	
CONS LANs	yes
WANs	yes
CLNS LANs	yes
WANs	yes
Transport layer	
0	yes
1	—
2	yes
3	—
4	yes
Session layer	yes
Presentation layer	planned
Message Handling (MHS) X.400	yes
Office Document Architecture (ODA)	—
Electronic Data Interchange (EDI)	—

SUPPLIER: GPT (GEC Plessey Telecom)

	Product Availability	Comments on Functionality
	System DATA SYSTEMS	
Architecture	?	GENERAL STATEMENT OF POLICY GPT Business Systems is committed to the early implementation of OSI products and to providing existing customers with a smooth migration path from existing proprietory protocols.
Interfaces		
V.24/V.28	yes	
X.21	yes	
X.35	yes	
V.36	yes	Where appropriate we will follow the
G.703	yes	recommendations of the GOSIP specifications. We have been active participants in the GOSIP workshops
WANs		and believe that these recommendations
X.25 (1980)	yes	are very significant in clarifying the
X.25 (1984)	yes	OSI implementation profiles. The close co-operation between UK GOSIP and other similar bodies encourages the
LANs		view that these recommendations will
8802/3 -	yes	be widely adopted by government,
8802/4 -	-	commercial and other procurement
8802/5 -	-	agencies.
LLC1	yes	
LLC2	planned 88/89	
N/W Service		
CONs LANs	planned 88	
WANs	planned 88	
CLNS LANs	yes	
WANs	-	
Transport layer		
0	yes	
1	-	
2	yes	
3	-	
4	yes	
Session layer	planned 88	
Presentation layer	-	
Message Handling (MHS) X.400	planned 88	
Office Document Architecture (ODA)	-	
Electronic Data Interchange (EDI)	-	

SUPPLIER: HEWLETT PACKARD

	Product Availability	Comments on Functionality
	System HP 9000	
Architecture		
WANs		
X.25 (1980)	yes	
X.25 (1984)	1st Qtr 89	
LANs		
8802/3 -	yes	
8802/4 -	yes*	* through 8802.3/8802.4 bridges
8802/5 -	support Q289	
LLC1	yes	
LLC2	yes	
N/W Service		
CONS LANs	-	
WANs	-	
CLNS LANs	yes	
WANs	yes	
Transport layer		
0	yes	
1	-	
2	yes	
3	-	
4	part product	
Session layer	yes	
Presentation layer	planned	
Message Handlings (MHS) X.400	yes	
Office Document Architecture (ODA)	-	
Electronic Data Interchange (EDI)	-	

SUPPLIER: HONEYWELL BULL (BULL)

	Product Availability			Comments on Functionality
	System DPS6	*System* DPS7/8	*System* XPS100	
Architecture	DSA	DSA	UNIX	
WANs				
X.25 (1980)	yes	yes	yes	
X.25 (1984)	1988	1988	1989	
LANs				
8802/3 -	yes	1989	yes	
8802/4 -	special prod.			
8802/5 -	1989			
LLC1	yes	1989	1988	
LLC2				
N/W Service				
CONS LANs	1988	1989	1989	
WANs	1988	1989	1988	
CLNS LANs	1988	1989	1988	
WANs	1988	1989	1989	
Transport layer				
0	1988	1989	1988	
1	-	-	-	
2	1988	1989	1988	
3	1988	1989	1989	
4	1988	1989	1988	
Session layer	1988	1989	1988	
Presentation layer	initial 1988	initial 1989	initial 1988	Dates reflect general release in UK
Message Handling (MHS) X.400 (84)	planned 1988	planned 89/planned	planned 1988	
Office Document Architecture (ODA)	planned	planned	planned	
Electronic Data Interchange (EDI)	planned	planned	planned	

SUPPLIER: IBM UNITED KINGDOM

	Product Availability	Comments on Functionality
	System /370	
Architecture	SNA	
WANs		
X.25 (1980)	yes	
X.25 (1984)	yes	
LANs		
8802/3 -	special product	Could be made available in specific cases
8802/4 -	planned	
8802/5 -	yes	
LLC1	yes	
LLC2	yes	
N/W Service		
CONS LANs	-	
WANs	-	
CLNS LANs	-	
WANs	yes	
Transport layer		
0	yes	
1	-	
2	yes	
3	-	
4	-	
Session layer	yes	
Presentation layer		Bridge to applications
Message Handling (MHS) X.400	announced	Available across /370 product range
Office Document Architecture (ODA)	-	
Electronic Data Interchange (EDI)	-	

SUPPLIER: INTERNATIONAL COMPUTERS

	Product Availability			Comments on Functionality
	System S39/2900*	*System* Clan	*System* DRS	*Not all upper layer (S) software is available on 2900 due to different VME release applicability Details provided for S39
Architecture	IPA	IPA	IPA	
WANs				
X.25 (1980)	yes	yes	yes	
X.25 (1984)	planned 88	planned 88	planned 88	ENV 41104
LANs				
8802/3 -	yes	yes	yes	ENV 41101 - ENV 41102
8802/4 -	-	special product	special product	
8802/5 -	-	-	-	
LLC1	yes	yes	yes	special product (gateway)
LLC2	-	-	-	
N/W Service				
CONS LANs	special product	special product	special product	
WANs	planned 88	planned 88	planned 88	
CLNS LANs	**yes	yes	yes	**current product to ENV 41101 supports iniative sub set of 8473 ENV 41102 in 90
WANs	-	special product	special product	
Transport layer				
0	yes	yes	yes	
1	-	-	-	
2	yes	yes	yes	
3	yes	yes	yes	
4	yes	yes	yes	
Session layer	part product planned 88	part product planned 88	part product planned 88	Not normally offered separately from layers 6 & 7
Presentation layer	-	-	-	
+ACSE	89	88	88	
Message Handling (MHS) X.400	part product planned, special product now	part product now	part product now	Clan and DRS current product based on ENV 41201/2 enhancements to full GOSIP conformance in 89. VME currently based on ECMA 93 –> X.400 (88) in 90. Converter products to ODA Standards are available on Clan and DRS (via ESPRIT PODA project). Subject to demand, general products are expected to appear 89/90 on S39/Clan/DRS.
Office Document Architecture (ODA)	-	special product 88	special product 88	
Electronic Data Interchange (EDI)	-	-	-	

SUPPLIER: INFORMATION TECHNOLOGY PLC (ITL)

	Product Availability
	System Momentum 9-10000
Architecture	One
WANs X.25 (1980)	yes
X.25 (1984)	-
LANs 8802/3 -	yes
8802/4 -	
8802/5 -	part range
LLC1	yes
LLC2	-
N/W Service CONS LANs	-
WANs	yes
CLNS LANs	yes
WANs	-
Transport layer 0	yes
1	-
2	-
3	yes
4	yes
Session layer	-
Presentation layer	-
Message Handling	yes
(MHS) X.400	
Office Document Architecture (ODA)	-
Electronic Data Interchange (EDI)	-

SUPPLIER: LOGICA UK

	Product Availability	Comments on Functionality
	System KERNELS	
Architecture	Portable	Product range called CPLEX.400 implementing X.400 (1988)

WANs
  X.25 (1980)
  X.25 (1984)
	
	yes
	-	

LANs
  8802/3
  8802/4
  8802/5
	
	-
	-
	-	

LLC1 LLC2	- -	

N/W Service
  CONS LANs
           WANs
  CLNS LANs
           WANs
	
	-
	-
	-
	-	

Transport layer
  0
 	
  1 -
  2 -
  3 -
  4 -
	class 0, 2, 4
	part product range
	3rd Qtr 88	

Session layer	part product range 3rd Qtr 88	BAS and BCS 1988 compatible
Presentation layer	PTS kernel now	X.409 functionality provided as a kernel: 'presentation transfer Syntax' available as product in its own right
Message Handling (MHS) X.400	planned mid 88	X.400 1988 model includes message store entity
Office Document Architecture (ODA)	-	
Electronic Data Interchange (EDI)	-	X.400 used as basic carrier for EDI

SUPPLIER: McDONNELL DOUGLAS

	Product Availability	Comments on Functionality
	System Series 19 (1)	(1) Series 19 is the largest range of Reality machines
Architecture		

WANs
  X.25 (1980)
  X.25 (1984)
	available
	available	

LANs
  8802/3   -
  8802/4   -
  8802/5   -
	available
	under consideration
	no plans	

LLC1 LLC2	available under consideration	

N/W Service
  CONS   LANs
            WANs
  CLNS   LANs
            WANs
	under consideration
	available
	available
	no plans	

Transport layer
  0
  1
  2
  3
  4
	planned 4th Qtr 88
	no plans
	planned 4th Qtr 88
	no plans
	available	

Session layer	planned 4th Qtr 88	
Presentation layer	planned 4th Qtr 88 (2)	(2) Series 19 will not give direct user access to lower layer services, ie access is via application layer
Message Handling (MHS) X.400	planned 4th Qtr 88	
Office Document Architecture (ODA)	under consideration	
Electronic Data Interchange (EDI)	under consideration	

SUPPLIER: NCR

	Product Availability	Comments on Functionality
	System Tower (UNIX)	
Architecture	SIA	
WANs 　X.25 (1980) 　X.25 (1984)	 yes planned	
LANs 　8802/3　　- 　8802/4　　- 　8802/5　　-	 yes special product planned 3rd Qtr 88	
LLC1 　LLC2	yes planned	
N/W Service 　CONS　　LANs 　　　　　WANs 　CLNS　　LANs 　　　　　WANs	 planned planned planned planned	
Transport layer 　0 　1 　2 　3 　4	 planned 3rd Qtr 88 - planned 3rd Qtr 88 special product planned 3rd Qtr 88	
Session layer	planned 3rd Qtr 88	
Presentation layer	planned	
Message Handling (MHS) X.400	planned 3rd Qtr 88	Full CEN/CENELEC compatible
Office Document Architecture (ODA)	-	
Electronic Data Interchange (EDI)	-	

Supplier: NORSK DATA

	Product Availability		
	System ND-100	*System* ND-500	*System* ND-5000
Architecture	COSMOS	COSMOS	COSMOS
WANs			
X.25 (1980)	yes	yes	yes
X.25 (1984	-	-	-
LANs			
8802/3	yes	yes	yes
8802/4	-	-	-
8802/5	-	-	-
LLC1	yes	yes	yes
LLC2	-	-	-
N/W Service			
CONS LANs	-	-	-
WANs	-	-	-
CLNS LANs	planned	planned	planned
WANs-	-		
Transport layer			
0	yes	-	yes
1	-	-	-
2	-	-	-
3	yes	-	yes
4	-	-	-
Session layer	planned	planned	planned
Presentation layer	planned	-	planned
Message Handling (MHS) X.400	planned	3rd Qtr 88 release	3rd Qtr 88 release
Office Document	-	planned	planned
Architecture (ODA)			
Electronic Data Interchange (EDI)	-	-	-

SUPPLIER: PRIME COMPUTER (UK)

	Product Availability		Comments on Functionality
	System 50 Series	*System* EXL Series	
Architecture	Primenet	-	NB.50 Series Primos op. sys. EXL Series UNIX op. sys.
WANs			
X.25 (1980)	yes	planned 88	
X.25 (1984)	yes	planned 88	
LANs			
8802/3 -	yes	yes	
8802/4 -	planned	-	
8802/5 -	-	-	
LLC1	planned	-	
LLC2	yes	planned	
N/W Service			
CONS LANs	yes	planned	
WANs	yes	planned	
CLNS LANs	planned	-	
WANs	-	-	
Transport layer			
0	planned	planned	
1	-	-	
2	planned	planned	
3	-	-	
4	planned	planned	
Session layer	planned	planned	
Presentation layer	planned	planned	
Message Handling (MHS) X.400	planned	planned	
Office Document Architecture (ODA)	-		
Electronic Data Interchange (EDI)	-		

SUPPLIER: SYDNEY COMMUNICATIONS

	Product Availability	Comments on Functionality
	System Messenger 400	
Architecture	portable	
WANs X.25 (1980) X.25 (1984)	yes part product range	Supports ISO 8208
LANs 8802/3 -	-	The products are portable s/w but have been run over 8802/3 LANS
8802/4 -	-	
8802/5 -	-	
LLC1 LLC2	yes yes	
N/W Service CONS LANs WANs CLNS LANs WANS	yes yes yes planned 88	Full protocol. Non-segmenting subset
Transport layer 0 1 2 3 4	yes yes yes yes yes	Supports CONS and CLNS (TP4 only) Max TPDU size from 128 to 8K octets
Session layer	yes	All functional units
Presentation layer	planned 88	
Message Handling (MHS) X.400	yes	P1, P2, RTS MMI proprietary directory X.400-1988 and Message Store planned
Office Document Architecture (ODA)	-	
Electronic Data Interchange (EDI)	planned 88	

SUPPLIER: TANDEM COMPUTERS EUROPE INCORPORATED

	Product Availability
	System NONSTOP
Architecture	EXPAND
WANs	
X.25 (1980)	yes
X.25 (1984)	some features available now others planned
LANs	
8802/3 -	planned
8802/4 -	planned
8802/5 -	-
LLC1	planned
LLC2	-
N/W Service	
CONS LANs	-
WANs	yes
CLNS LANs	planned
WANs	-
Transport layer	
0	yes
1	yes
2	yes
3	yes
4	planned
Session layer	planned
Presentation layer	planned
Message Handling (MHS) X.400	planned
Office Document Architecture (ODA)	- -
Electronic Data Interchange (EDI)	some support planned

SUPPLIER: UNISYS

	Product Availability				Comments on Functionality
	System A Series	*System* B2X/3X	*System* 1100/2200	*System* UNIX	
Architecture	BNA	BTOS/BNA B Mail= sub architecture	DCA	DCA	
WANs X.25 (1980) X.25 (1984)	yes planned 89	yes planned 89	yes planned 89	yes planned	
LANs 8802/3 - 8802/4 - 8802/5 -	planned 88 planned -	yes planned -	3rd Qtr 88 planned -	yes planned -	
LLC1 LLC2	planned 88 -	yes -	3rd Qtr 88 -	yes -	
N/W Service CONS LANs WANs CLNS LANs WANs	- - planned planned	- planned 88 planned planned	- - planned planned	- planned 88 planned planned	availability depends upon transport class
Transport layer 0 1 2 3 4	planned 89 - planned 89 - planned	3rd Qtr 88 - 3rd Qtr 88 - planned 88	3rd Qtr 88 - 3rd Qtr 88 - planned	2nd Qtr 88 - 2nd Qtr 88 - planned 88	NCC conformance tested on 1100
Session layer	planned 89	3rd Qtr 88	3rd Qtr 88	2nd Qtr 88	
Presentation layer	planned 89	planned 88	planned 88	planned 88	These dates are for full presentation Elements of presentation needs for MHS/FTAM are included in those products
Message Handling (MHS) X.400	planned 89	3rd Qtr 88	3rd Qtr 88	2nd Qtr 88	CEN/CENELEC and NBS Conformance
Office Document Architecture (ODA)	planned	planned	planned	planned	
Electronic Data Interchange (EDI)	-	-	-	-	Awaiting OSI clarification, participating in UK GOSIP Committees

SUPPLIER: WANG (UK)

	Product Availability	Comments on Functionality
	System VS	
Architecture	WSN	

WANs
  X.25 (1980)
  X.25 (1984)
	yes
	planned
	(Q1/89)	

LANs
  8802/3   -
  8802/4   -
  8802/5   -
	yes
	-
	to be
	announced 4/88	

LLC1 LLC2	to be announced 6/88 ,, ,,	

N/W Service
  CONS   LANs
 	
          WANs
  CLNS   LANs
          WANs
	to be
	announced 6/88
	   ,, ,,
	   ,, ,,
	   ,, ,,	

Transport layer
  0
 	
  1
  2
 	
  3
  4
	to be
	announced 6/88
	-
	to be
	announced 6/88
	   ,, ,,
	   ,, ,,	

Session layer BAS	avail 6/88	
BSS BCS	,, ,, ,, ,,	To be announced 6/6/88. This is a gateway to the Wang OFFICE proprietary messaging system and will support PRMD and ADMD connections and act as a full MTA within a private domain. The user interface will be Wang OFFICE. Wang OFFICE directory services and Freeform addressing will provide X.400 addressing capabilities.
Presentation layer ASN.1	planned 89	
Message Handling (MHS) X.400 1984	to be announced 6/88	Generally available late 88 *: via Wang OFFICE
Office Document Architecture (ODA) WITA/ODIF	planned 89	A filter for Wang Information Transfer Architecture to ODIF
Electronic Data Interchange (EDI) EDIFACT	planned 89	Not necessarily OSI transported initially

Appendix 9
ISO and CCITT Communications Standards

INTRODUCTION

This appendix attempts to list some of the areas of OSI standardisation being undertaken by both ISO and the CCITT. It does not constitute a definitive list, but rather an attempt to illustrate the variety and complexity of the available standards.

CCITT Recommendation	ISO/IEC Standard or Technical Report
	Architectural Models
X.200	ISO 7498, OSI Basic Reference Model (1984)
	ISO 7498/AD 1, Connectionless Mode Transmission (1987)
	ISO 7498/DAD 2, Multi-peer Transmission
	ISO 7498/2, Security Architecture
	ISO 7498/3, Naming and Addressing
	DIS 7498/4, Management Framework
	DP 9545, Application Layer Structure
	Descriptive Conventions
X.208	ISO 8824, Specification of Abstract Syntax Notation One (ASN.1)(1987)
	ISO 8824/AD 1, Specification of Abstract Syntax Notation One (ASN.1) – Addendum 1: Covering Real, Subtypes, etc
X.209	ISO 8825, Specification of the Basic Encoding Rules for Abstract Syntax Notation One (ASN.1) (1987)
	ISO 8825/AD 1, Specification of the Basic Encoding Rules for Abstract Syntax Notation One (ASN.1) – Addendum 1: Covering Real, etc
X.210	ISO TR 8509, OSI Service Conventions (1987)
X.407	ISO 10021/3, MOTIS-Abstract Service Definition Conventions
	Conformance
X.290	DP 9646/1, OSI Conformance Testing Methodology and Framework – Part 1, General Concepts
	DP 9646/2, OSI Conformance Testing Methodology and Framework – Part 2, Abstract Test Suite Specification

CCITT Recommendation	**ISO/IEC Standard or Technical Report**
	DP 9646/3, OSI Conformance Testing Methodology and Framework – Part 3, The Tree and Tabular Combined Notation
	DP 9646/4, OSI Conformance Testing Methodology and Framework – Part 4, Test Realisation
	DP 9646/5, OSI Conformance Testing Methodology and Framework – Part 5, Requirements on Test Laboratories and their Clients in the Conformance Assessment Process
	Common Application Service Elements
X.217	ISO 8649, Service Definition for Association Control Service Element (ACSE)
X.227	ISO 8650, Protocol Specification for ACSE
	DIS 9804, Commitment, Concurrency and Recovery (CCR)
	DIS 9805, Protocol Specification for CCR
X.218	ISO 9066/1, Reliable Transfer – Part1, Model and Service Definition
X.228	ISO 9066/2, Reliable Transfer – Part 2, Protocol Specification
X.219	ISO 9072/1, Remote Operations – Part 1, Model, Notation and Service Definition
X.229	ISO 9072/2, Remote Operations – Part 2, Protocol Specification
	Peer-to-Peer Application
	ISO 8571/1, File Transfer, Access and Management – Part 1, General Description
	ISO 8571/2, File Transfer, Access and Management – Part 2, Virtual Filestore Definition
	ISO 8571/3, File Transfer, Access and Management – Part 3, File Service Definition
	ISO 8571/4, File Transfer, Access and Management – Part 4, File Protocol Specification
	DP 8571/5, File Transfer, Access and Management – Part 5, Protocol Im-plementation Conformance Statement (PICS Proforma)
	DIS 9040, Virtual Terminal Basic Class Service
	DIS 9041, Virtual Terminal Basic Class Protocol
	DP 9579, Remote Database Access (RDA)
	Multi-party Applications
	DP 10026/1, Distributed Transaction Processing – Part 1, Model
	DP 10026/2, Distributed Transaction Processing – Part 2, Service Definition
	DP 10026/3, Distributed Transaction Processing – Part 3, Protocol Specification
	DIS 8831, Job Transfer and Manipulation (JTM) Concepts and Services
	DIS 8832, Job Transfer and Manipulation (JTM), Specification of the Basic Class Protocol for JTM

CCITT Recommendation	ISO/IEC Standard or Technical Report
	Distributed Applications
X.500	ISO 9594/1, The Directory – Overview of Concepts, Models and Services
X.501	ISO 9594/2, The Directory – Models
X.511	ISO 9594/3, The Directory – Abstract Service Definition
X.518	ISO 9594/4, The Directory – Procedures for Distributed Operation
X.519	ISO 9594/5, The Directory – Protocol Specification
X.520	ISO 9594/6, The Directory – Selected Attribute Types
X.521	ISO 9594/7, The Directory – Selected Object Classes
X.509	ISO 9594/8, The Directory – Authentication Framework
	DP 9595/2, Management Information Service Definition – Part 2, Common Management Information Service (CMIS) Definition
	DP 9596/2, Management Information Protocol Specifications – Part 2, Common Management Information Protocol Specification
X.400	ISO 10021/1, Message Oriented Text Interchange System (MOTIS) – Systems and Service Overview
X.402	ISO 10021/2, MOTIS – Overall Architecture
X.407	ISO 10021/3, MOTIS – Abstract Service Definition Conventions
X.411	ISO 10021/4, MOTIS – Message Transfer System: Abstract Service Definition and Procedures
X.413	ISO 10021/5, MOTIS – Message Store: Abstract Service Definition
X.419	ISO 10021/6, MOTIS – Protocol Specification
X.420	ISO 10021/7, MOTIS – Interpersonal Messaging Systems
	DP 10031/1, Distributed Office Applications Model (DOAM) – Part 1, General Model
	DP 10031/2, Distributed Office Application Model (DOAM) – Part 2, Referenced Data Transfer
	Presentation Layer
	DIS 8822, Connection-oriented Presentation Service
	DIS 8823, Connection-oriented Presentation Protocol
	Session Layer
X.215	ISO 8326, Connection-oriented Session Service
X.225	ISO 8327, Connection-oriented Session Protocol
	Transport Layer
X.214	ISO 8072, Transport Service Definition
X.224	ISO 8073, Connection-oriented Transport Protocol
	ISO 8073/ADD 1, Network Connection Management Sub-protocol

CCITT Recommendation	**ISO/IEC Standard or Technical Report**
	Network Layer
	ISO 8208, X.25 Packet Level Protocol
X.213	ISO 8348, Network Service Definition
	ISO 8348/ADD 1, Connectionless Network Service (CLNS)
	ISO 8348/ADD 2, Network Layer Addressing
	ISO 8473, Connectionless Network Protocol
	DP 8648, Internal Organisation of the Network Layer
	ISO 8878, Use of X.25 to provide Connection-oriented Network Service (CONS)
	DIS 8880/n, Provision of Network Service
	DIS 8881, Use of X.25 Packet Level Protocol in LANs
	Data Link Layer
	ISO 7776, HDLC: X.25 LAPB Compatible DTE Data Link Procedure
	DIS 8802/2, Logical Link Control for LANs
	Physical Layer
X.21	DTE/DCE Physical Level Interface Characteristics
LANs (effectively include data link functions)	
	ISO 8802/3, CSMA/CD Technology
	ISO 8802/3 DAD 1, CSMA/CD Reduced Geography (Cheapernet)
	ISO 8802/4, Token Bus Technology
	ISO 8802/5, Token Ring Technology
	Physical to Network Layer
X.25	Access to Packet Switched Networks

Appendix 10

NIST Implementor's Agreements for X.400 Additions to Chapter 8, Appendix B

INTRODUCTION AND SCOPE

This is a guideline for EDI data transfer in an X.400 environment conforming to the NIST agreeements. These recommended practices outline procedures for use in transferring EDI transactions between trading partner applications in an attempt to facilitate actual X.400 implementations by EDI users.

The scope of this guideline is to describe specific recommendations for adopting X.400 as the data transfer mechanism between EDI applications.

Model

The MHS recommendations can accommodate EDI through the approach illustrated in Figure A.10.1. Many Message Transfer (MT) service elements defined in the X.400 recommendations are particulary

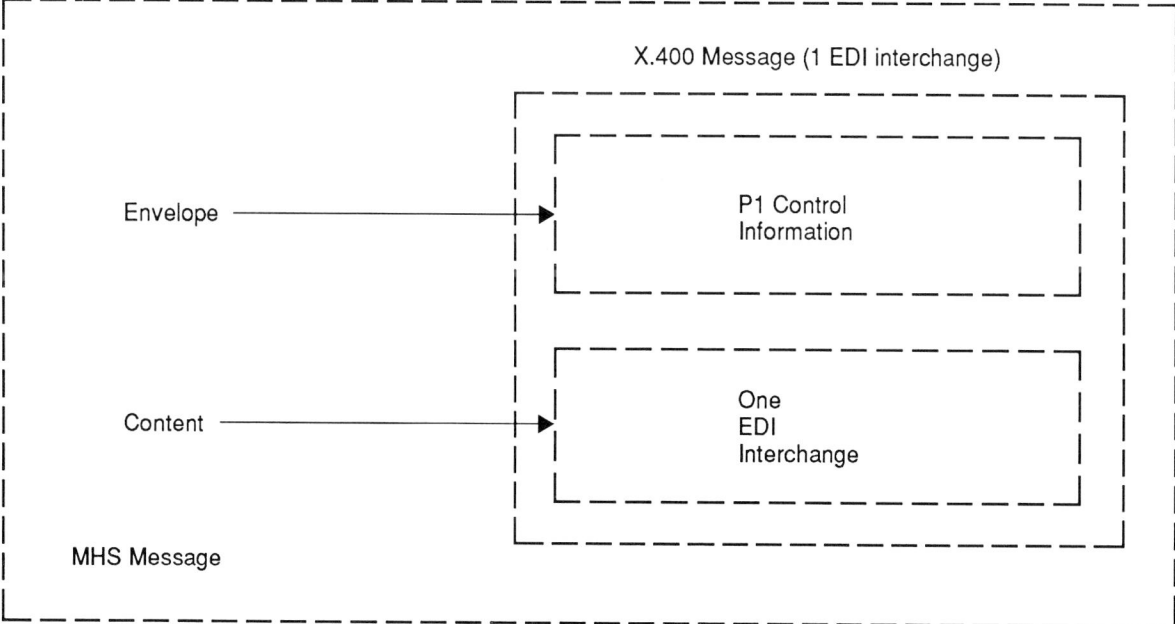

Figure A.10.1

229

useful to the EDI application. This diagram depicts an EDI content (1 EDI interchange) envelope by the P1 MHS envelope.

All the MT services defined in the X.400 Recommendations may be used for EDI. However, it is not required to support optional or non-essential services exchange EDI data between EDI users.

When an EDI user submits an EDI Trade Document to the EDI User Agent, the EDI UA will submit the EDI content plus P1 envelope to the Message Transfer System (MTS) (see Figure A.10.2).

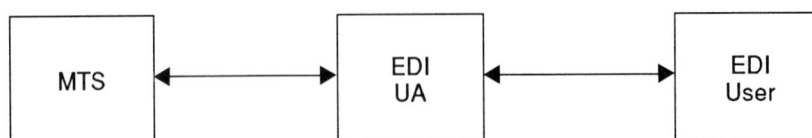

Figure A.10.2

The EDI UA must support the essential MT Services as defined in these Agreements; for example, as a minimum, to provide default values for services not elected by the EDI user, such as Grade of Delivery.

Note: MT Services are not necessarily made available by the EDI UA to the EDI user.

Protocol Elements Supported for EDI

The following P1 protocol elements will be used to support EDI applications:

Content Type

For EDI applications, the content type will be 0 (undefined content).

Original Encoded Information Types

Any EIT defined in the X.400 Recommendations may be used to specify the encoding of EDI content. However, for ANSI X12 EDI applications in particular, it is expected that the 'undefined' and 'Ia5Text' EIT's will normally be used, with 'undefined' used to signify the EBCDIC character set.

Addressing and Routeing

It is anticipated that connection of some existing systems to an X.400 service for EDI purposes will be by other than X.400 protocols, at least in the short term.

EDI messages entering the X.400 environment will therefore need to have X.400 O/R Names added to identify the originating and recipient trading partners, typically by means of local directory services in the originating domain which will map EDI identifiers/addresses into O/R Names. Such O/R Names will contain Standard Attributes as defined in Table 8-12 and for recipient trading partners will at least identify the destination domain.

In the case of trading partners outside the X.400 environment it is expected, however, that there will be cases where message delivery will require the provision of addressing information beyond that which can be carried in Standard Attributes. In such cases, Domain Defined Attributes are recommended to be used.

The syntax of this DDA is as defined in Table 8-12, with a single occurrence having the type name 'EDI' (upper case) and a value containing the identifier/address of the trading partner. For ANSI X12 purposes, specifically, this value will comprise the 2-digit interchange ID qualifier followed by the interchange ID (max. 15 characters).

Routeing on this DDA shall only occur, if at all, in the destination domain.

Addition to Table 8-12

Note 5: Some existing systems which will be accessed via an X.400 service (whether directly connected using X.400 protocols or otherwise) may require the provision of addressing attributes which are not adequately supported by Standard Attributes as defined in these Agreements. In such cases, Domain Defined Attributes are an acceptable means of carrying additional addressing information. Failure to support the specification and relaying of DDAs may prevent successful interworking with such existing systems until such time as all systems are capable of relaying and delivery using only the Standard Attribute list. Specific recommendations on the use of DDAs for particular applications are in the Recommended Practices section.

Index